HISTORY of BUILDING

HISTORY of BUILDING

styles, methods and materials

George Balcombe

Batsford Academic and Educational
London

About the book

The 20 buildings described show how the styles, methods and materials of building have changed since Stonehenge was put up nearly 5000 years ago. All these buildings still exist in some form and they are all in the United Kingdom, although further examples are given of buildings in Europe and the United States of America.

The book has five parts. Part 1 **Stone Circles** looks at the earliest structures and at Roman building in Britain. The Anglo-Saxon style leads into Part 2 **Pointed Arches**, where the Gothic style and its revival in Victorian times tell their own story.

The Golden Rectangles of Part 3 are a clue to the secrets of the Classical style and of buildings such as St Paul's Cathedral in London. Part 4 is called **Iron Rules** because nearly 200 years ago iron, and then steel, took over from wood and stone as the main structural material and, over many years, rules for using the metal materials were worked out by tests and mathematics. Part 5 brings the story of styles, methods and materials into the industrial world of the 1980s where **High Tech** is not only the technology of high buildings but also of the microchip.

First published 1985

Typeset by Keyspools Ltd, Golborne, Lancs
and printed in Great Britain by
Anchor-Brendon Ltd
Tiptree, Essex

for the publishers
Batsford Academic and Educational
an imprint of B T Batsford Ltd
4 Fitzhardinge Street
London W1H 0AH

British Library Cataloguing in Publication Data

Balcombe, George
History of building: styles, methods and materials
1. Architecture—Great Britain—History
I. Title
720'.941 NA961

ISBN 0 7134 2186 X

PART 1 STONE CIRCLES

1 Stonehenge

Place Salisbury Plain, 14·4 km (9 miles) north of Salisbury, Wiltshire
Time *Start* Stage 1, 2800 BC; Stage 2, 2100 BC; Stage 3, 2000 BC
Finish after 1100 BC
Construction time It is not known how long each stage took, but it would have been many years
Purpose Not known. But almost certainly a place for rites and for observing the sun and moon
Style Post and beam

As well as Stonehenge other huge circles of standing stones were built on and near Salisbury Plain. At Avebury, 38 km (24 miles) north of Stonehenge, is Europe's largest stone circle. Its outside diameter is 427 m (1400 ft). The biggest stone weighs over 40 tonnes (40 tons). Prehistoric sites crowd the high chalk downs between Stonehenge and Avebury. (*Figure 1.1.*)

Of the few people living in the south of England in prehistoric times, most were farmers or hunters and they kept to the safe high ground above the dangerous forests and marshes. Several racial groups came and mingled and settled. Trade in gold from Ireland, and in

1.1 Looking south to the stone circles. The Avenue leads off on the left

1.2 Trilithons. The nearest one is part of the inner U. The others are part of the outer circle

amber from Scandinavia, passed to and fro across Salisbury Plain. The ruling people were rich, and built many large sites to do with magic rites and spirits of the dead, including hundreds of special mounded graves called **barrows**. Wiltshire alone has over 2000 round barrows, and more than 80 long barrows such as the 107 m (350 ft) long barrow at West Kennet. The people also built round camps – perhaps as corrals for cattle. There were also flint mines and race tracks, one of which, the Cursus, measures nearly 3 km (2 miles) long.

So Stonehenge was only one among thousands of sites. But it took longer to build than any other on Salisbury Plain, and some people believe that it was never finished and that as seen today it may bear little resemblance to what its builders intended. (*Figure 1.2.*)

What we see at Stonehenge today is the ruin of Stage 3 which began about 2000 BC. The huge standing stones and the fallen ones are the remains of several stone circles built around the same centre. (*Figure 1.3.*)

The method of construction used is called **post-and-beam**. Two stone uprights carry a stone beam, and the three huge stones together make a **trilithon**. The ring of trilithons at Stonehenge is unique among prehistoric structures all over the world. (*Figure 1.4.*)

A thousand years after Stage 1 at Stonehenge, the Egyptians were building trilithon temples along the River Nile which were bigger and stronger than the Stonehenge construction. By 1400 BC the Greeks were building with vast stones at Mycenae, and by 500 BC the three-stone trilithon had become the Classical style in Greece. The two stone uprights had become round columns, and the stone beam had become the **entablature**. A Classical entablature is a beam made up of several layers of stone fitted and slotted together.

Stonehenge was built as a circle of trilithons. Inside the circle stood five taller and more massive trilithons in a U shape. Some of the stone beams have now fallen, likewise some of the stone uprights; others stand alone; some are missing, and other stones, half buried, litter the site. All these stones are inside an even greater circle made by a ditch and a bank. The outside diameter of the bank is 97·5 m (320 ft).

A single entrance led into the circle from the north-east. A 3 km (2 mile) long road leaves the circle and winds gently down the slopes to the River Avon. This road, called the Avenue, was the last part of Stonehenge to be built, sometime after 1100 BC.

The bluestones, of which Stonehenge was partly constructed, were thought to possess magical properties so they were dragged down from the Prescelly mountains in Wales to the sea and taken some 216 km (135 miles) by boat up the Bristol Channel and then by river and land routes to Stonehenge. (*Figure 1.5.*)

Hundreds of years elapsed between each of the three main stages of building at the site.

Stage 1: begun about 2800 BC

This consisted of the ditch and bank, also the Aubrey Holes, which comprised a ring of 56 holes in the ground just inside the circle. The holes contained human remains, suggesting human sacrifices. A single upright stone, the Heel Stone, 4·5 m (15 ft) high, was erected outside the entrance.

The ditch was dug to a diameter of 116 m (380 ft) and the earth from it was piled up to make the bank. The entrance faced north-east. The New Stone Age builders fashioned their tools by sharpening sticks and hardening the ends in a fire. With these digging sticks they removed the turf. They then levered chunks of chalk out of the ground with picks shaped from the antlers of red deer. About 80 of their antler picks (*figure 1.6*) have since been found in the ditch. The ditch diggers also used small shovels made from ox shoulder blades. Three hundred men at a time would have been needed to dig the ditch and build the bank and six to seven hundred men to drag the Heel Stone weighing 35·5 tonnes into place.

Stage 2: begun about 2100 BC

The bluestones were arranged in two concentric circles. The 80 bluestones, each weighing 4 tonnes, were transported from the Prescelly mountains in the south-west tip of Wales in six stages:

1 Down 48 km (30 miles) from the mountains to the sea at Milford Haven. Forty men would have been needed to drag one bluestone.

2 The stones were then taken separately by boat or raft, perhaps in convoy for safety, to the mouth of the River Severn near Bristol.

(In 1954 BBC Television staged a reconstruction of this with four boys and a 4 tonnes concrete block on three purpose-built punts fixed side by side. They transported the block on water to demonstrate that the bluestones could, in good weather conditions, have been taken by sea.)

3 By boat up the River Avon at Bristol to about 11 km (7 miles) beyond Bath, then up the River Frome.

4 Forty men would then have dragged each bluestone, by sledge, 9·6 km (6 miles) overland from Frome to Warminster.

5 Down the River Wylye to Salisbury. Then a sharp turn north up Hampshire's River Avon to Amesbury.

6 More teams of 40 men dragged the bluestones from the river uphill for 3 km (2 miles) to Stonehenge.

In spite of all this effort Stage 2 of Stonehenge was never finished. The reason why remains a mystery.

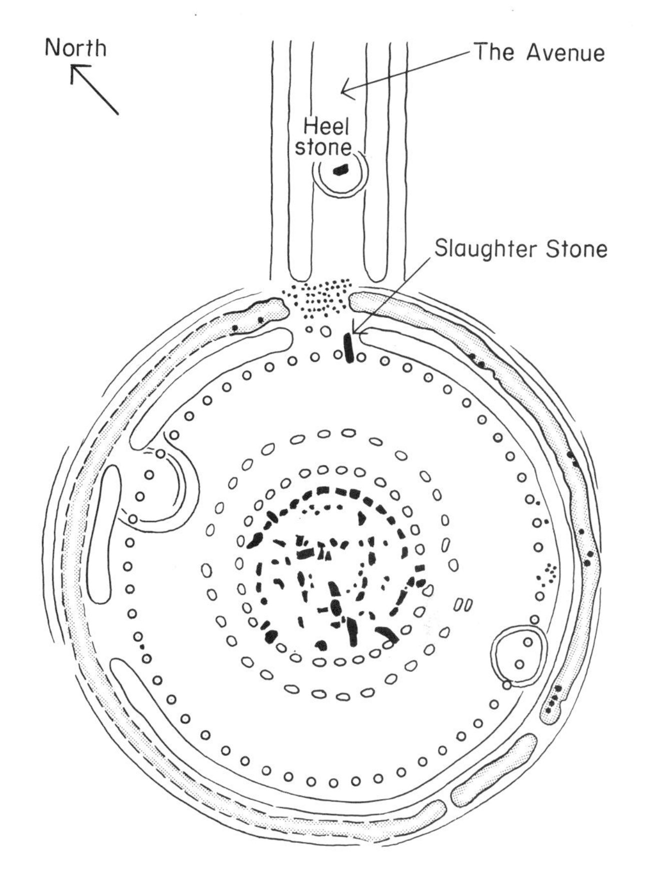

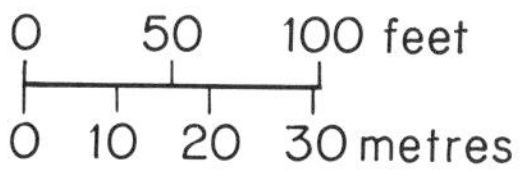

1.4 Plan

1.3 Looking north-west from the air. The Avenue is on the right. The ditch, bank and Aubrey Holes can be seen

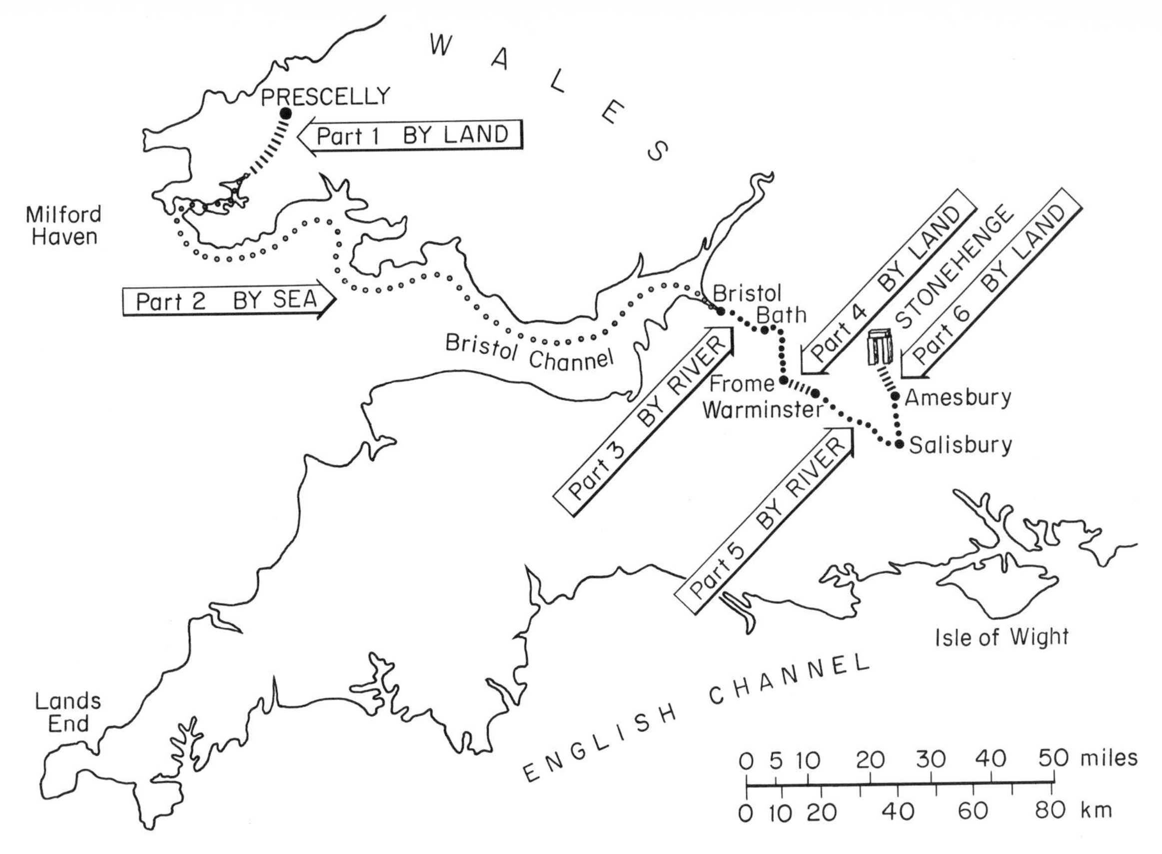

1.5 Route map of bluestone transport
1.6 Antler pick

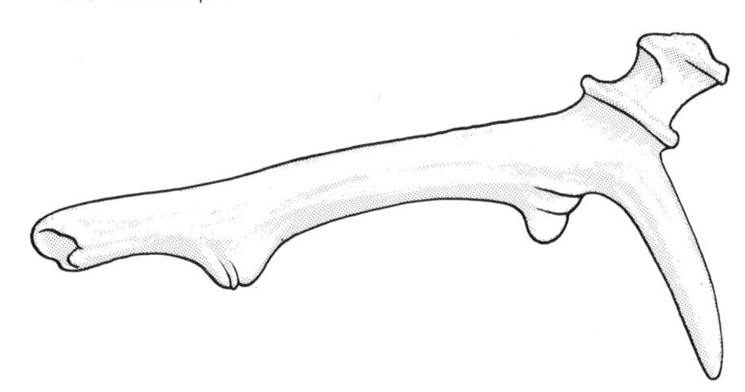

Stage 3: begun about 2000 BC
In this period were introduced the huge stones from Avebury called **sarsens**. The first bluestones were taken down and the stones used again later. The sarsen circle and the U shape of sarsen trilithons were erected.

The 81 huge sarsen blocks were dragged to Stonehenge 38 km (24 miles) south from the Marlborough Downs near Avebury (*figure 1.7*). Each stone weighed up to 50·8 tonnes; 1500 men would have taken nearly six years – ten according to some experts – to transport the stones to the site. Each trip from Avebury to Stonehenge and back again took over two months. The giant sarsens were transported in heavy wooden cradles on rollers, with twisted hide thongs being used as ropes.

The complete circle of trilithons was built with the smaller stones to a diameter of 29·5 m (97 ft). Bigger stones were used for the U shape inside the circle.

The huge stone uprights were erected by being tipped over the edge of a hole, and men then pulled them upright by means of hide ropes. The bottom of each sarsen was slightly pointed so that it could be rocked from side to side until exactly positioned. One of the long sarsens had 2·4 m (8 ft) buried in the ground so, of its total length of 8·8 m (29 ft), only 6·4 m (21 ft) was above ground. This stone required nearly 200 men on the ropes. Friction stakes were driven into the chalk sides of the hole to prevent the uprights, when tilted, from crushing them.

Then the 6·35 tonnes stone beams had to be lifted. The Stonehenge builders did not have block-and-tackle, pulleys or cranes, so would most likely have used levers in the following way: one end of the beam would be lifted by lever and a block of wood slipped underneath, and similarly with the other end of the beam. This process would be repeated with more blocks. The beam would rise slowly a few centimetres (inches) at a time and gradually a platform of criss-cross logs was built up for the levering to go on. Ropes fixed to the end of the lever were pulled by men on the ground.

Each lift required seven men pulling down on a 4·25 m (14 ft) long lever with the fulcrum 304 mm (1 ft)

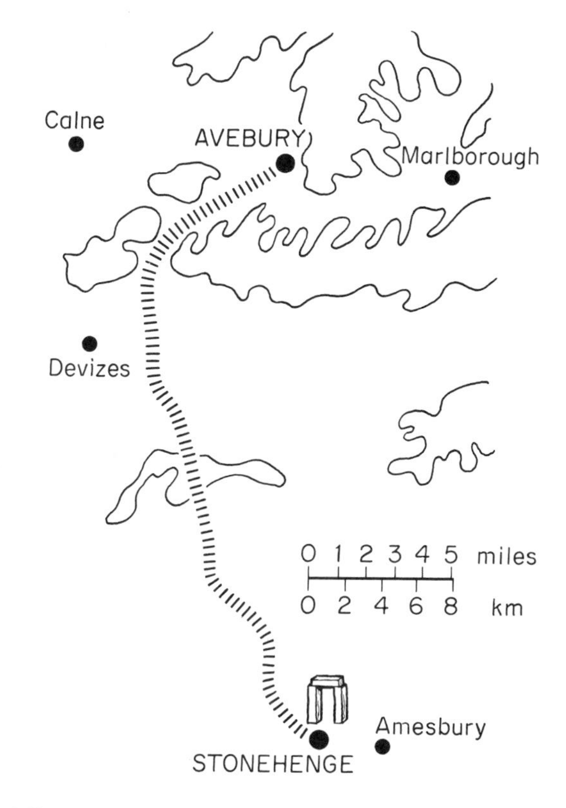

1.7 Route map of sarsen transport

from the end of the lintel. When the beam at last reached the right height, levers were again used to swing the lintel sideways onto the top of the two uprights. Each upright had a knob on the top – a **tenon** – which fitted into a hole under the beam.

The huge sarsens were taken out of the ground as thick diamond-shaped slabs of sandstone, close to the surface of the Marlborough Downs. They had been formed over thousands of years from an even bed of sand, so their two faces were parallel surfaces smoothed by nature. The builders of the stone circles at Avebury left many of their sarsens in the natural diamond shape, but the Stonehenge builders wanted the upright slabs to have straight sides and so the slabs had to be split. Sandstone splits easily at right-angles to the flat surface of the slab.

Two methods could have been used to shape the Stonehenge uprights:

Method 1 When the sarsen slab had a natural crack in the appropriate place, the builders **pecked** at the crack with stone tools. As the crack widened and deepened, wooden wedges were driven into it, water poured in, and the wedges began to swell. The crack opened more and more until the stone split.

Method 2 When no natural crack existed in the sarsen slab, a crack had to be created. Fires were lit on the top face of the slab along the desired line. The sudden cooling of the hot stone by water set up stresses inside the stone. A row of men along the line then simultaneously smashed heavy stone hammers, called **mauls**, on the sarsen, thus causing the slab to split. However, if the blows from these football-sized mauls only made a slight crack in the sarsen, the first method would be used to complete the split.

The sarsens may have been made roughly to size before starting their 38 km (24 mile) journey south across Salisbury Plain to Stonehenge. When they reached the site nearly six weeks later the sarsen slabs had to be 'dressed'. **Dressing** is the work done by the mason to make a rough stone block into the right size and shape. The stone surface is also tooled to make the right planes.

Sarsen is a harder stone than granite and among today's tools only the hardest alloys of steel can dent or smooth its surface. Since the Stonehenge masons had no metal tools they pounded the sarsens with their stone mauls. Many of these mauls have been found on the site 4000 years later, some weighing up to 27 kilos (60 lb).

The masons dressed the stones by several processes:

1 Mauls weighing 13·5 kilos (30 lb) made grooves 228 mm × 76 mm (9 in. × 3 in.) along the sarsen's length.

2 Mauls weighing 27 kilos (60 lb) then smashed the ridges between the grooves.

3 A fine dressing followed with smaller mauls making grooves 228 mm (9 in.) long and 50 mm (2 in.) wide and 6 mm (0·25 in.) deep. Whole sides were dressed by this method.

4 Mauls flattened all remaining grooves and ridges, leaving an even and pitted surface.

5 Ropes pulled heavy grinders made of sandstone to and fro over the sarsen slabs. An abrasive such as crushed flint mixed with water gave the hard sarsen surface a smooth finish.

It took many years to dress and polish all the sarsens at Stonehenge. The rough dressing alone would have taken three years if 50 masons worked seven days a week and ten hours a day all the year through. Perhaps the first stones to reach Stonehenge were just being polished when the last stones were transported to the site from Avebury six years later.

The fine polishing was mainly done on the sides of the stones which would have been visible from inside the stone circles. The biggest U-shaped trilithon of all, situated in the middle of the circle, was polished on both sides. This suggests that the rites at Stonehenge were performed inside rather than outside the stone circles. It is not known now what these rites were; the builders and their secrets are buried under the hundreds of barrows on Salisbury Plain all round Stonehenge.

2 Skara Brae

Place Bay of Skail, Orkney Islands, Scotland
Time *Start* and *finish* about 2100 BC. But the site was built on and altered three times over hundreds of years
Construction time It is not known how long each stage took
Purpose Houses, or perhaps some sort of hostel
Style New Stone Age or Neolithic

2.1 A house at Skara Brae. The fireplace is in the middle. And from left to right the built-in stone fittings are wall cupboards, a stone bunk-bed and a wall-unit or sideboard. In the background is the seashore

The rooms seem to have been built for a special group of people who were part of a race living mostly in Scotland and the North of England though some may have settled as far south as the River Thames. The Skara Brae people were farmers who bred sheep and long-horned cattle. The grass in this area was good so they could keep all their herds in one district. They grew no corn and although they ate seafood such as limpets found on the shore, it seems they were not fishermen.

Of all the New Stone Age houses found so far, those at Skara Brae are the most complete and provide the best illustration of Stone Age buildings and how the people lived in them. (*Figure 2.1.*)

In 1866 a storm blew away sand dunes on the shore at Skara Brae and this was how the houses were discovered. They had been wrecked by another storm, around 2000 BC, and the people had built these houses in hollows of the sand dunes right by the sea. Huddled low and close together the houses gave some shelter from Atlantic gales.

There were eight buildings with narrow passages twisting between them which were roofed and paved. Proper drains were laid under the houses. Each house had one room only. The corners were rounded and the floor level was below general ground level outside. The rooms ranged in size from 6 m × 5·5 m to 4·25 m × 3·6 m (20 ft × 18 ft to 14 ft × 12 ft). So these New Stone Age rooms were much the same size as many rooms in houses today. (*Figure 2.2.*)

The rooms at Skara Brae had built-in furniture. As there were no trees on the island, thin slabs of stone were used to make the built-in fittings. One room had a 2 m (6·5 ft) bunk bed of stone and two smaller ones, perhaps for children. The people slept on heather mattresses. Animal skins hanging from stone bed-posts gave extra warmth and privacy. Shelves were built-in above the beds. The builders made other storage places such as cupboards and boxes, and containers for liquid. Heat came from peat burning on a rectangular hearth in the middle of the room. The door of the house was only 1·2 m (4 ft) high and less than 610 mm (2 ft) wide.

The low walls of the house were built with thin and flat stones laid in uneven courses without mortar. Thicknesses varied being in some places 1·5 m (5 ft) and in others 457 mm (1·5 ft). Some of the inside stones were carved with zigzag or triangular patterns. There was no plaster on the walls. As there were no windows, daylight may have filtered into the room through a smoke hole in the roof.

It is not known how the roof was made but the stone courses at the top of the walls narrowed inwards. This **corbelling** reduced the roof span. Rare pieces of drift-wood or whale-ribs may have been collected for the roof frame. This frame would have rested on the stone corbelling and then have been covered with grass as is the method still used today in some old farms in Norway and Iceland.

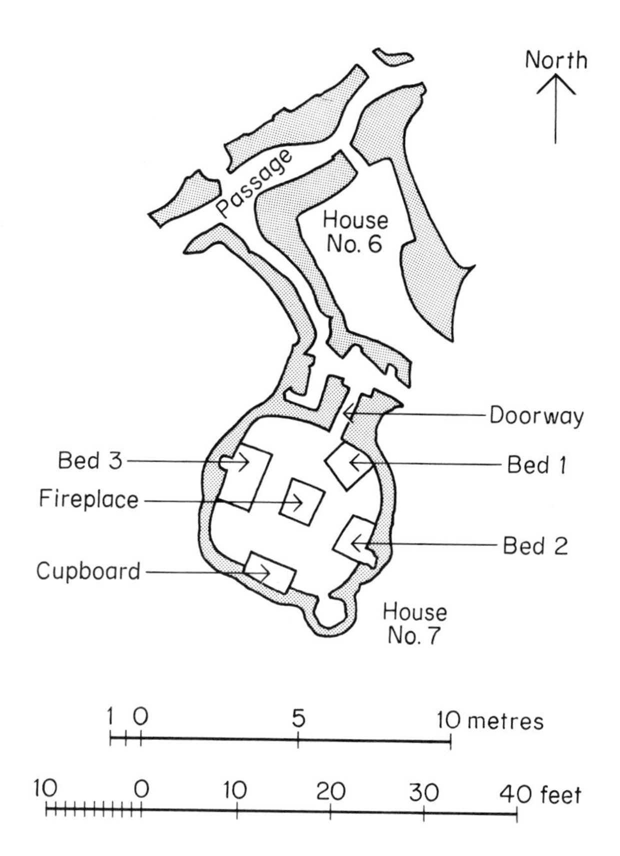

2.2 Plan of houses No. 6 and No. 7. Roofed passages wind between them. All the shapes are irregular

The Skara Brae people wore leather clothes and made all their own tools and household things, as well as jewellery made out of bone. When they fled from the great storm in 2000 BC someone dropped a trail of beads in their hurry. Sand slowly drifted over the ruins, houses and beads staying hidden for 4000 years.

3 Fishbourne

Place Fishbourne, near Chichester, Sussex
Time *Start* about AD 75.
Finish about AD 80. Many later alterations
Construction time 5 years
Purpose A district headquarters. Sometimes called a 'palace'
Style Classical. The Roman style in Britain

The Roman Army took control of the Sussex coast in AD 43. They built a harbour at Fishbourne for ships to land troops and supplies. By AD 75 they were working on a huge group of buildings close to the harbour for use as a headquarters. These one-storey buildings were spread over 4 hectares (10 acres). The main block measured 152·5 m × 100·6 m (500 ft × 330 ft) and the garden in the middle was 97·6 m × 76·25 m (320 ft × 250 ft). It is the only large Roman garden so far found north of the Alps. (*Figure 3.2.*)

3.1 Boy on a dolphin. Part of a mosaic floor designed and laid by Italian craftsmen

The main block held law courts and offices and guest rooms. There were dining rooms and a big kitchen area. There were also rooms for small swimming pools and for something like today's saunas. Fishbourne's top man also had a big suite of rooms for himself and his family. They lived in the south wing and had another large garden which went down to the harbour. (*Figure 3.3.*)

West from the main block lay the many workshops and stores and houses for the people whose work kept the harbour and headquarters going. There were also farm buildings and stores to house the produce from the farms stretching inland from the site.

The style used by Roman designers is called **Classical**. There were two main periods when the Classical style was used.

Period 1: 1000 years
From 500 BC to AD 500. The Greeks invented the Classical style and it was in full use by 500 BC. Then the Romans began to use it soon after 200 BC but the Roman world was in ruins by AD 500.
Period 2: 500 years
From AD 1450 to 1950. Italian designers were the first to start using the Classical style again. They made scale drawings of Roman ruins and found out Roman methods of design which they used for their own buildings. Their rediscovery of the Classical style is called the **Renaissance** which means 're-birth'. The Classical style spread from Italy to France by the early 1500s and to Britain by the early 1600s. By the 1950s almost no designers were using the Classical style for new buildings. By the 1980s nearly all building work in the Classical style in Britain was for one-off private houses and the conservation of older buildings or for sites among existing Classical buildings.

There were three main design methods used for Classical buildings.

Design method 1: Axis lines

These were straight lines used for laying out towns and buildings. Streets and open spaces in towns were laid out along **axis lines** and the buildings were placed so that the axis line went right through them in the middle. (*Figures 3.2 and 3.4.*)

Sometimes one axis would cross another at right-angles and where they crossed there would be a main open space outdoors or a main room indoors. Designers also made the building's features balance like a mirror image each side of the axis.

At Fishbourne the main axis line was over 1·6 km (1 mile) long. A road was built along the axis line to link the harbour and headquarters with the growing town at Chichester. The axis line went right through the middle of Fishbourne's main buildings and garden. Two of the main rooms were placed on the axis line and so was the garden's main path with its fountains. (*Figures 3.3 and 3.4.*)

Design method 2: Grid lines

These were straight lines like the network of lines on a map and helped to place one thing in relation to another.

Classical designers used grid lines to make layouts for whole towns and buildings, and to fix the height of rooms together with the position and shapes of doors and windows. Designers liked to use rows of columns and they placed each column on a grid line.

In stone buildings every pair of columns carried a beam made from a single stone. So the length of stones that could be quarried fixed the distance between the columns.

This distance was then divided into parts called **modules**. For example 8 modules was usual. Two modules then became the diameter of the column. The height of the column was then fixed at 18 modules. The module itself was divided into 30 parts which became the unit of measurement for all the details of the whole building. (*Figure 3.8.*)

Classical designers therefore measured their buildings by modules not by metres or feet. Modules fixed the distance between columns and their diameter and their height. The columns governed the whole building becoming, along with their entablature beam, the basic unit in the Classical kit-of-parts.

A column-and-beam was usually one storey high but sometimes it was two storeys. Two or sometimes three rows of columns would be placed on top of one another.

The network of grid lines and module measurements set up a system which helped designers to fix the position and size of all the building's rooms and features. This system is called **symmetry** and Classical designers had to learn how to use it.

At Fishbourne there were rows of columns on all four sides of the garden. The grid lines were spaced at 3·35 m

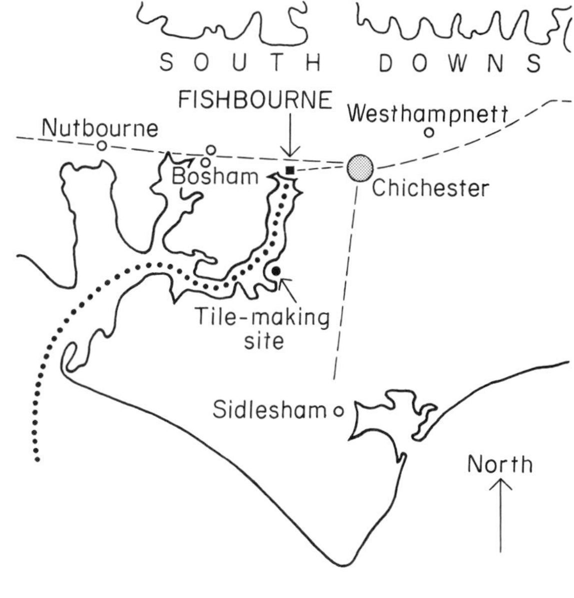

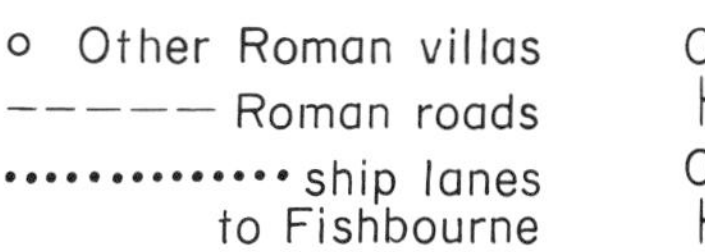

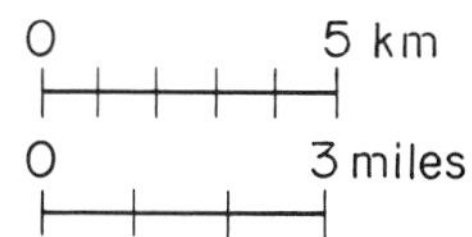

3.2 Map of Fishbourne area

3.3 Site plan of the Roman headquarters

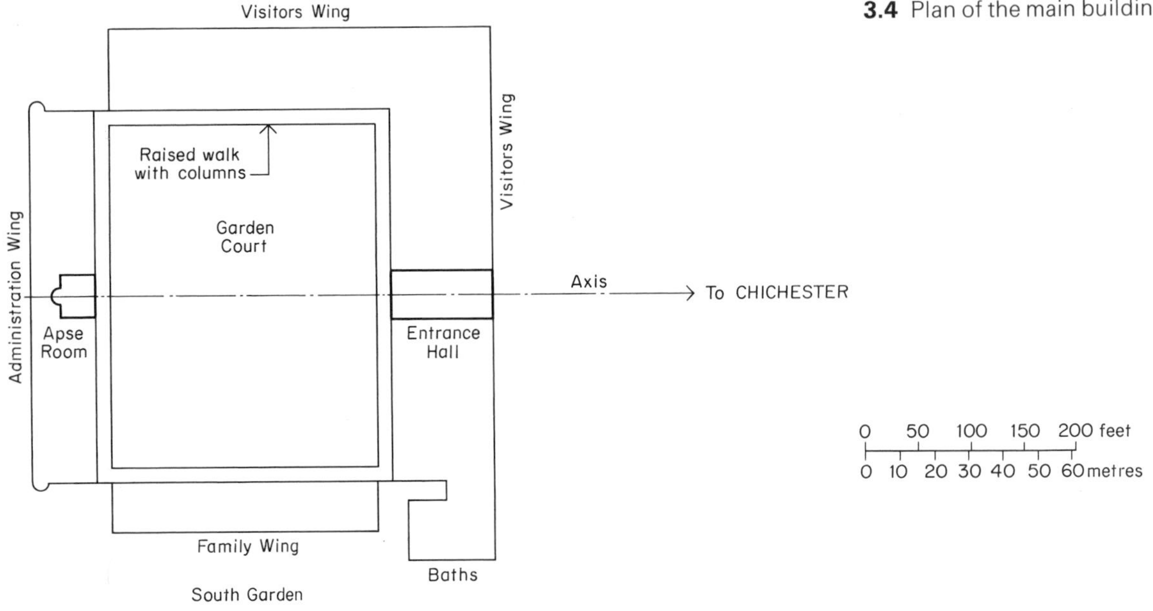

3.4 Plan of the main building

(11 ft) with a column on each line. Other columns at Fishbourne were placed on each side of the axis line. The 33 m (108 ft) long entrance hall had a porch of six columns at each end which were 8 m (26 ft) high. Across the other side of the garden was a main room on the axis line. This room had a 6 m (20 ft) diameter recess called an **apse** at the end. People entered this apse through a porch of four columns.

A porch with columns is called a **portico** and it often carries a roof gable called a **pediment**. A portico with a pediment is one of the main features of the Classical style and designers often placed porticos on axis lines. It was a method of showing where an entrance was and of making the entrance look important.

Classical designers also used grid lines in another way. The outside and inside walls of a Classical building are designed on a grid of rectangles. This grid fixes the position and shape of wall openings such as doors and windows, and the position and shape of the solid wall between openings. The grid also fixes the size and position for any rows or pairs of columns-and-beam placed in front of a wall.

The diagonals of the rectangles were frequently used in the grid. For example the diagonal going across the front of a building from a top corner to a bottom corner was parallel to the slope of the gable pediment over the portico. (*Figure 3.5.*)

The rectangle which Classical designers found most useful was called the **Golden Rectangle**. This had sides in the ratio of 1 to 0·618. For practical use this often became a rectangle with the short side measuring five units and the long side eight units. Geometrically the Golden Rectangle comes from the side of a square and the diagonal on half of that square.

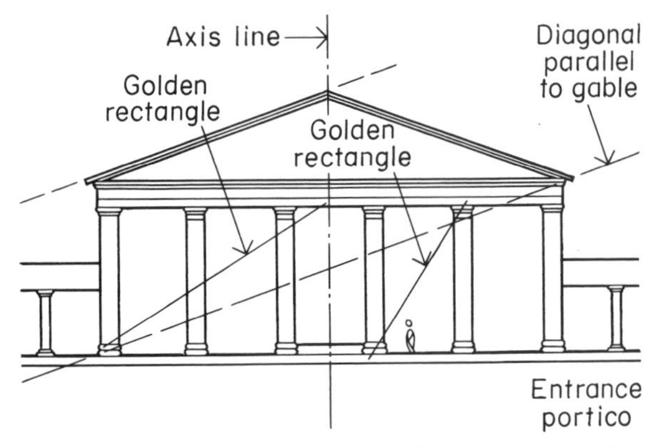

3.5 The main entrance portico with six columns and a pediment. Axis line in the centre

At Fishbourne the portico design was based on rectangles and their diagonals. (*Figures 3.6 and 3.7.*)

Design method 3: Outlines

Classical designers liked to shape the outlines of their buildings. Any part of the building which was to be seen in profile such as the outside corners, the rooftop or the edges of wall openings for doors and windows, was given special shapes.

They also made special shapes when two different features of a building joined together. For example the top or **capital** of a column was carved where it carried the stone entablature beam.

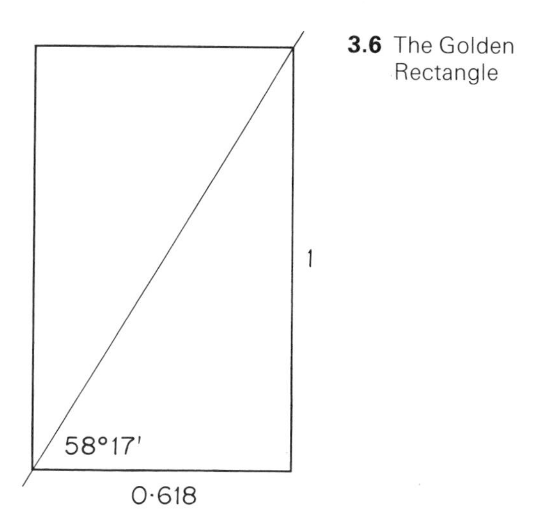

3.6 The Golden Rectangle

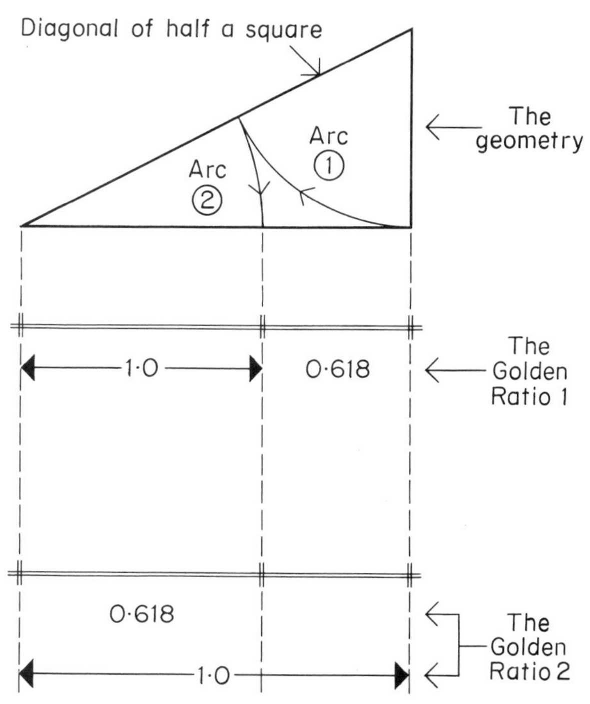

3.7 The geometry of the Golden Section

By 500 BC the Greeks had invented the column-and-beam as a main feature in the Classical designer's kit-of-parts. The column-and-beam was called an **Order**. (*Figure 3.8.*) They could be used with the columns standing free of the wall or with them built-in as part of the wall. Built-in columns were called **pilasters**. Free columns tapered towards the top but pilasters had straight sides and were flat instead of round. Sometimes half-round columns were built into walls.

Among the many people who had their own ideas about the Orders was the Scottish designer William Chambers. He was born in Sweden in 1723 but built large projects in and around London, such as Somerset House by the River Thames, begun in 1776. As a teenager he had been a sailor, visiting Africa, India and China. In 1758 he designed the ten-tiered Chinese pagoda in London's Kew Gardens.

But William Chambers's real love was for the Classical style which he studied first in Paris and then in Rome from 1750 to 1755. He thought there were six Orders from which designers should work and he described as follows their sizes and details. (*Figure 3.8.*)

1 **The Doric Order** The Greeks invented this and were using it by 500 BC. Chambers gave the column height as 12 modules and the entablature height as 4 modules. The column's capital was a saucer-shaped disc.
2 **The Tuscan Order** The Romans invented this and were using it by AD 70 at Rome's vast arena, the Colosseum. William Chambers gave the column height as 14 modules and the entablature as 3·5 modules. Its capital was also a disc but it had a base pad as well.
3 **The Roman Doric Order** An even thinner column at 16 modules high with a 4 module entablature. There was a disc base as well as a capital.
4 **The Ionic Order** The Greeks invented this and were using it by 550 BC. The Romans also used it. William Chambers thought its column should be a slim 18 modules high and its entablature 4·5 modules. The capital was carved as two spiral shapes like wheels. Spirals had been carved on buildings since at least the time when Stonehenge's trilithons went up. On the island of Malta the Temple of Tarxien was carved with spirals in about 2000 BC. The spirals were symbols to make people think about life after death.
5 **Corinthian Order** The Greeks invented this and were using it by 420 BC. Later the Romans used it more than the Greeks. Chambers thought its column should be the thinnest of all the Orders at 20 modules high with 5 modules for the entablature. The column's capital was carved as leaves of the acanthus plant.
6 **Composite Order** The Romans invented this and were using it by AD 81. Chambers showed its sizes as the same as the Corinthian Order's, but its capital had Ionic spirals on the top half and Corinthian leaves on the bottom half.

There were other versions of the Orders as well as Chambers's six. All of them needed skilled designers and skilled craftsmen to make them in marble, stone or wood. At Fishbourne a version of the Roman Tuscan Order was used. It was the simplest Order of all to design and build.

Three main methods of construction were used for Classical buildings: walls; arches and vaults; column-and-beam.

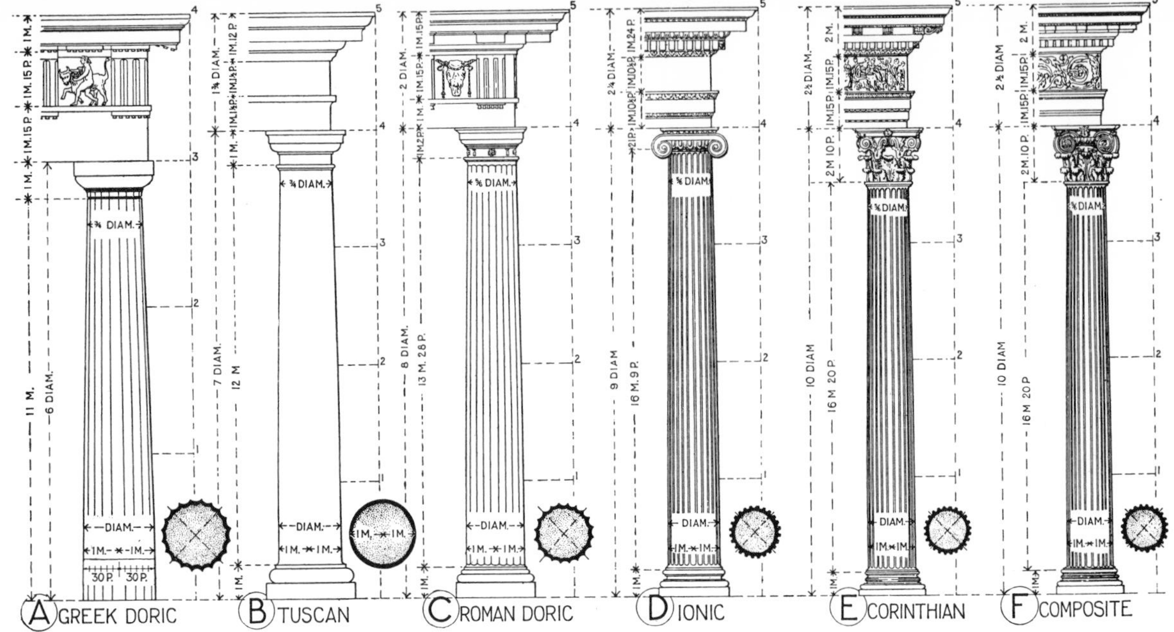

3.8 Sir William Chambers' six Orders

Construction method 1: walls

These were basic to Classical design. They were load-bearing and carried other walls, floors and roofs. Walls took the outward thrust of arches and vaults. Walls divided the building's inside spaces, and provided large surfaces on which the designer could get to work with his grids of rectangles and diagonals. Many Classical walls had Orders in front of them. Windows, doors and arched openings often went into the rectangle of wall between columns and their entablature. Where the outside wall of a ground floor was built of stone the joints were made to look big and the stones were cut roughly. This **rustication** made the building look strong. Rustication was another main feature of the Classical style.

Unlike the house builders at Skara Brae, the Greeks and Romans built walls with great precision. The Greeks used huge blocks of stone fitted tightly together without mortar. By 200 BC the Romans were using lime mortar and building with mass concrete. Many of the largest and most richly decorated buildings in Rome had cores of concrete which was poured in between parallel walls of brick or stone. The concrete walls of the round Pantheon temple built in Rome between AD 120 and 124 were 6 m (20 ft) thick.

Fishbourne had nothing so massive. Its stone walls varied in thickness from 457 mm to 610 mm (18 in. to 24 in.). To level the 275 m × 122 m (900 ft × 400 ft) site about 27,000 cu m (36,000 cu. yd) of clay and gravel had to be dug out and carried away. The foundation trenches were 762 mm (2·5 ft) wide. Oak piles 150 mm (6 in.) in diameter were rammed into the bottom of the trenches. Each room needed from about 120 to 150 such piles. Then the concrete footings were laid ready for the stone walls which were built with squared blocks of stone set in lime mortar. The highest walls were 7·5 m (25 ft) above floor level.

The floors were built of concrete and many were finished with **mosaics**. Mosaics were coloured patterns and pictures made with 6 mm (0·25 in.) square fragments of marble and tile called **tesserae**. Skilled craftsmen from Italy laid Fishbourne's mosaic floors between AD 75 and 80. The most well-known floor shows a boy riding a dolphin. (Figure 3.1.)

Polished marble slabs up to 19 mm (0·75 in.) thick lined some of Fishbourne's inside walls. Artists decorated other walls in bright colours with paints which they mixed themselves and applied when the top coat of plaster was still wet. These floors and walls were a luxury as were the cast glass panes in the windows and the rich textiles used for furnishings.

Construction method 2: arches and vaults

Greek designers seldom built arches. They used single stone beams to carry walls across openings. But the Romans put up huge half-circle arches. These arches could be built with small stones which were easy to quarry and take to a site. Arches could span wider openings than single stone beams but they pushed outwards at the bottom so thick walls or buttresses had to be placed there to resist this push.

The Romans also built two kinds of half-round stone roofs called **vaults**: barrel vaults and cross vaults. **Barrel vaults** were half cylinders of stone or brick rising from the top of the walls, so the vault was the same width as the room beneath it. The half-round ends of the vault were open and windows could be put there. No other openings in the vault were practical as they made it weak. Barrel vaults also pushed outwards at the bottom and needed walls or buttresses.

Cross vaults are also known as **groin vaults**. These were made from two barrel vaults which crossed each other at right angles. This made four open ends. The groin was the line where the half-cylinder surfaces of the two barrel vaults met. The groin was a half ellipse. In each cross vault two groins crossed at right angles and they were the weakest part of a cross vault. Many vaults collapsed along the lines of the groins. Cross vaults could only be built over a square space, but several such squares with a vault could be built in a row side by side.

At Fishbourne the only structural stone arches and barrel vaults were those built over the swimming pools and saunas. To heat the swimming pools and saunas hot air and gases from a furnace went through flues under the floor and then up through hollow tile blocks in the walls and so out into the open air.

This Roman heating system was called a **hypocaust**. Not only saunas in houses but, by AD 211, also large public swimming baths in Rome were heated by hypocaust systems. Hypocausts needed a great deal of both wood and labour. At Fishbourne another furnace heated water in a bronze boiler. A stone barrel vault which was fire-proof and steam-proof, was built over these saunas and a wood roof was built on top to keep off the weather.

The Fishbourne barrel vaults only spanned about 5 m (17 ft) from wall to wall. But in Rome cross vaults spanned up to 25 m (83 ft) and were built 37 m (120 ft) above floor level. The stones of arches and vaults were laid on semi-circular wooden framework called **centering**. This was removed when the half circles of stone were complete.

At Fishbourne the main room with its 6 m (20 ft) apse also had a barrel vault. But it was only a decorative one made of plaster on wood framing. Its diameter was 10 m (33 ft). Plaster vaults were later used in many rooms during the Classical style's second period from AD 1450 to 1950. The carpenter needed as much skill to make the framing for a plaster vault as to make the centering for a stone vault, and the same skill went into making roofs with wood trusses.

Construction method 3: column-and-beam

The Greeks were using this method by 500 BC. They often put rows of outdoor columns-and-beams on all four sides of a building. The Romans sometimes used columns at one end only as a portico. Some columns were made from one long stone but others were built up from separate stones called **drums**. The Greeks joined their marble column drums together at the centre with bronze dowels set in holes and fixed by molten lead. The surface joint between two drums was so fine that it could hardly be seen.

At Fishbourne various kinds of stone were used for the 3·65 m (12 ft) high columns around the garden. Limestone came from Bembridge on the Isle of Wight and from Gloucestershire. Stone was also shipped over from Caen in the north of France. Caen stone was to become a main building material for Gothic cathedrals over a thousand years later. And in AD 1681 Christopher Wren used it inside his Classical style St Paul's Cathedral in London.

The local people had never seen any building like Fishbourne. For over a thousand years before the great building went up near the new harbour in AD 75 the Celtic farmers and traders of Britain had lived in the roughest of buildings. They grouped some of their round houses on hill tops behind wide security circles of ditches and banks. The largest kinds of such hill forts enclosed up to 20 hectares (50 acres). Other groups of houses were built down by rivers where cattle could graze in lush water-meadows. Some house groups were fenced in for safety. Others without fences were more like villages along a street.

The Celtic people took quickly to the Roman way of life. The richer people with their gaudy clothes and heavy gold jewellery liked the Roman idea of luxury. They began to build rectangular houses. To keep up with the Roman invaders they put in saunas heated by hypocausts and got specialists in to lay mosaic floors. If they could afford only one mosaic floor it usually went into their dining-room which was on an axis line and had an apse at one end. Just like the palace at Fishbourne.

Houses of this sort were built on country estates and as in Italy were called **villas**. Most Roman villas in Britain were lived in by Britons or by European refugees fleeing invasions by Barbarians from the east. One villa at Winterton in Lincolnshire was built on top of the 15 m (50 ft) round Celtic house the owner's family used to live in. Most villas were built with a wood frame filled in with fencing and covered with mud or plaster. No villa was as vast or as richly decorated as the building at Fishbourne. Nothing but foundations and a few mosaic floors remain of these villas now.

Some time between AD 270 and 280 a fire destroyed Fishbourne. The site was abandoned and the building's stones were taken away for new buildings in Chichester. Soil slowly covered the site. It became farmland and stayed farmland until 1960 when a man using a mechanical digger for the drains of a new housing estate cut into a mass of roof tiles and mosaic fragments. He had found the ruins. The site was bought and a museum built. Today people can follow the story of the once-great headquarters and see its foundations. They can look down from raised walkways onto the mosaic floors which volunteers have slowly pieced together again. And visitors can wander in the garden now laid out and planted again just as it was 1700 years ago.

4 Earls Barton

Place Earls Barton, Northamptonshire
Time *Start* and *finish* between 1000 and 1050
Construction time Not known as there are no records
Purpose A church. The tower is now the main feature of the building
Style Anglo Saxon

After the Roman way of life collapsed, the Saxons invaded and settled in Britain during the AD 400s and 500s so becoming 'Anglo' Saxons. They became the next builders of main projects after the Romans but at best their buildings were crude. Almost nothing is left today of their work.

Most of the people were farmers and built simple houses and farm buildings with wood frames and wicker walls covered with mud. The roofs were thatched. Others lived in the ruins of Roman cities such as Canterbury, Lincoln, London or York and took building materials from the Roman ruins there. St Alban's Abbey in Hertfordshire, for example, was built with bricks from the Roman town of Verulamium.

Augustine, the Christian priest, brought the method of building walls with stones bedded in mortar when he came to Britain from Rome in AD 600. But mostly stone was used for the new churches springing up everywhere. The remains of a few can be seen today. They were built during the 400 years from AD 650 to 1050, though Danish invaders put a stop to all main projects between 800 and 950.

Churches were built by putting up thick stone walls. Some churches had wide openings inside made by arches, others only had door-size openings. Windows were small. Sometimes the walls were built with patterns outside and this was done on the tower at Earls Barton.

At first the Anglo-Saxon priests were interested in the style and methods used for new churches in Rome and other main centres in Europe. But after the Danish invasions they did not seem to care what was being done abroad. From 950 to 1050 the main concern of Anglo-Saxon designers was not construction but decoration. Money was spent on gold and silver objects and rich clothes for the priests to wear.

During the first 150 years of stone construction, from 650 to 800, several churches were built at Canterbury, three being built before 670. A church was built at

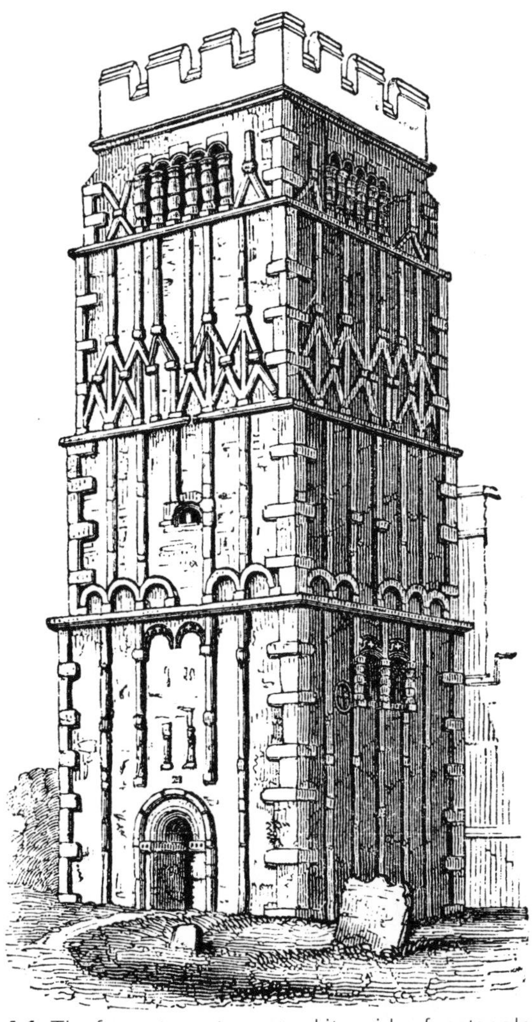

4.1 The four-storey tower and its grids of rectangles

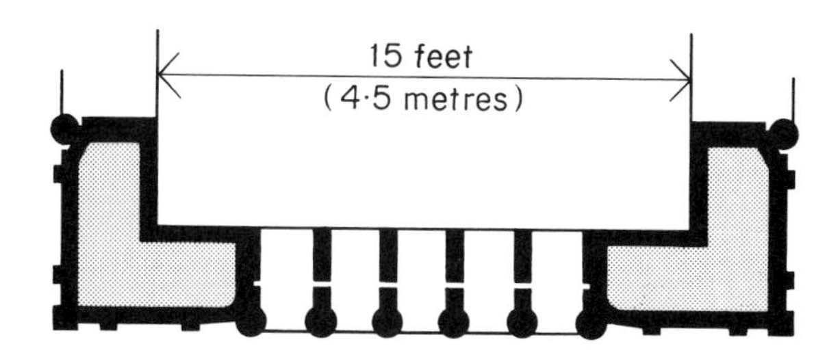

4.2 Plan of the top storey

Brixworth in 670, another in Abingdon in 680 and one in Glastonbury in 700. In the year 770 York Cathedral was rebuilt. But this was not the building seen at York today. The present cathedral was being built 500 years after the Anglo-Saxon one. The style at York in the 1270s was Gothic.

Between 950 and 1000 nearly 40 new church centres such as monasteries were set up by the Anglo-Saxons. In 999 a stone church with a tower at each end was built in Durham. There were others at Deerhurst, North Elmham and Ramsey. A round church went up at Canterbury, and at Repton in Derbyshire a church basement was built with spirally twisted columns. And in this active 50 years, before 1000, other and earlier church centres were brought to life again.

Near the end of this activity the church tower at Earls Barton was built, sometime between 1000 and 1050. The tower was 9 m (30 ft) square at the base and over 23 m (76 ft) high with four storeys each narrower than the one below. (*Figures 4.1 and 4.2.*)

Five outside features of the Earls Barton tower make it look as if its builders had heard about, but had not seen, Roman designs in the Classical style. Fishbourne had already been a ruin under farmland for 700 years by the time Earls Barton tower was built. Its features are as follows:

1 Grid of Rectangles These bands of stone run along the top of each storey and other thin bands of stone run upright between them. These divide the whole tower into a grid of rectangles.
2 Diagonals Other bands of stone are built as diagonals of the rectangles. And this also suggests that the Anglo-Saxon builders had heard of the way Roman designers used diagonals.
3 Half circles The Earls Barton builders put half circles of stone at the top of some of the rectangles, just like the Roman designers used half-circle arches in walls. The tower's doorway also had a half-circle top.
4 Corner stones Like Roman designers the builders at Earls Barton put thick stones called **quoins** at the corners. These quoins made the walls look even stronger.
5 Columns Windows in the tower were flanked by bulging and stumpy columns with stones carved to look like arches above their capitals. Again it seems the Anglo-Saxon designer had heard about, but had not seen, Classical columns.

Earls Barton tower was already standing in the year 1055 when Britain's last Anglo-Saxon king commissioned the building of a new Westminster Abbey in London by the River Thames. Again as at York, this is not the Abbey that thousands of people now visit every year. The present building was started by the 1250s and was in the Gothic style.

King Edward put his 1055 Abbey on the site of an earlier Anglo-Saxon church built in AD 616. But the king's Westminster Abbey was not designed in the Anglo-Saxon style. It was Norman. The designers got their up-to-date ideas from the new church in the north of France at Jumieges, begun about 15 years before in 1040.

Not much remains today of the last Anglo-Saxon king's Norman style at Westminster Abbey. But it shows how the Norman style dominated design in Britain 11 years before the invading Norman army in 1066 took over everything else.

4.3 Windows with stumpy columns

PART 2 POINTED ARCHES

5 Durham Cathedral

Place Durham
Time *Start* 1093
Finish 1133
Construction time 40 years
Purpose A church. When the senior priest is a bishop his main church is called a 'cathedral'
Style Norman. Also called English Romanesque

5.1 The cliff-top cathedral and its three towers

During the AD 900s Europe became rich and peaceful enough to have big buildings for its main centres of government and trade. Churches were centres of power and wealth too and the ever-increasing number of new churches showed this. They were large and simple. Geometry ruled the stone shapes and spaces and, by the 980s, a new style called **Romanesque** had evolved. In 981 a large church in the Romanesque style was built at Cluny in France and many others followed in the 990s. In the year 1000 a Romanesque church was built at Hildesheim in Germany.

The Normans took part in this church building boom. From their small but active area in northern France, called Normandy, the Normans had a big effect on life in Europe. Four years before their army landed in Britain the Normans built a big stone church at Caen in northern France. It was from quarries at Caen that the Romans had shipped stone to Fishbourne nearly a thousand years before.

Caen's Trinity church of 1062 was one of the first large Norman churches. In 1066 the Normans invaded Britain and in 1093 work began on the great cathedral at the northern town of Durham. In some ways it was similar to the one at Caen.

Like Roman buildings in the Classical style these Norman churches were laid out on axis lines and grid lines. At Durham a long axis line and a short one crossed at right angles. A tower was built over this crossing. A tall and narrow hall with lower side spaces was built along the axis line and a second hall was built along the short axis line. Both halls and their side spaces were also laid out on grid lines. (*Figure 5.4.*)

The east-west hall was over 113 m (370 ft) long and the north-south hall 67 m (219 ft) long. They were both 22·25 m (73 ft) high. With later additions the building's total length today is 143 m (470 ft).

Like walls built in the Classical style Durham's inside walls were divided into a grid of rectangles. The walls had three storeys, each with its own row of arches. Massive piers held the arches up. A stone vault covered the halls and it rested on top of the piers. The vault's outward push was resisted by buttress walls on half-circle arches built at right angles to the piers. The tall part of the hall was the **nave**. Each of the three storeys in the nave walls was different. (*Figures 5.2, 5.3 and 5.5.*)

1 Bottom storey

This was 12·5 m (41 ft) high. It opened into the low side spaces called **aisles** through half-circle arches in the 2·4 m (8 ft) thick walls. The openings through to the aisles were designed as a cluster of slender columns each carrying its own slender arch. On alternate grid lines a massive cylinder of stone carried the arches.

2 Middle storey

This was 5·8 m (19 ft) high. It opened into the **gallery** in the roof over the aisles. The vault buttresses were in

5.2 Looking east down the Norman nave. Pointed arches go from side to side. Segment arches go across as diagonals

the gallery. The openings in the wall at this level were also designed as clusters of columns. But under the half-circle arch two extra smaller arches were constructed side by side.

3 Top storey

This was 4 m (13 ft) high. It opened to the outside and was used for windows. At this level there were three narrow arches on small slender columns.

This scheme of nave walls in three storeys became a standard design for large churches during the next 450 years. (*Figure 5.6.*) The building programme stopped in the 1540s when King Henry VIII took over the Church and its property in England, so ending the wealth and power of the priests.

5.3 Half-circle arches in the Norman nave. The vault ribs have carved patterns

Norman designers and builders knew the methods of vaulting used a thousand years before by the Romans in their Classical buildings. The Normans built a barrel vault and cross vaults at the Tower of London between 1078 and 1098. But at Durham Cathedral the designers used another type of vault. This was the **rib** vault.

Barrel and cross vaults were built like eggshells of stone. But rib vaults were built of arches like the ribs of animals. The voids between these stone ribs were filled afterwards with surfaces of thin and small slabs of stone.

When a barrel or cross vault was being built its whole surface had to be supported by a wooden frame underneath. The wooden frame had to be built up from the floor. The method was slow and expensive. But when a rib vault was being built it only needed a wooden frame under the thin rib of stone. It was quick and cheap.

The rib vault was not invented by the Normans. It was in use 900 years before Durham Cathedral was built, not in Europe but in what today is the border area between Iran and Iraq. Precision skills were needed to build a rib vault. By the AD 600s Islamic designers had used rib vaults in many buildings around the Mediterannean region.

During the AD 800s and 900s there were masons and carpenters in Spain who had been trained in the Islamic methods of building. They could build with greater precision than any other craftsmen in Europe. A church was built in AD 913 at Leon in Spain by these Islamic methods.

The rib vaults which the Normans built at Durham between 1128 and 1133 were among the first ones in Europe. A cross vault could only be built over a square space, since the curves of the vault's surfaces fixed the shape of the groins. (See *figure 5.7*.) But a rib vault could be built over a rectangular space, the shape of the rib fixing the curves of the vault's surface. This simple but basic difference between a cross vault and a rib vault gave builders a new method of vaulting. It was to lead in the next 400 years to the beautiful and daring structures of the Gothic cathedrals. (*Figures 6.1 and 7.1.*)

In Norman rib vaults the ribs were half circles or segments of circles. This made them easy to build. But

5.4 Plan

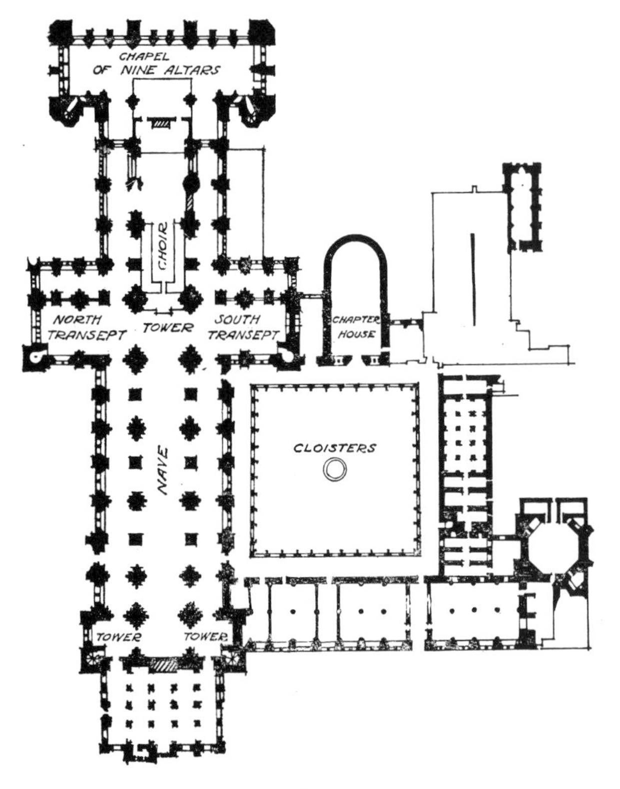

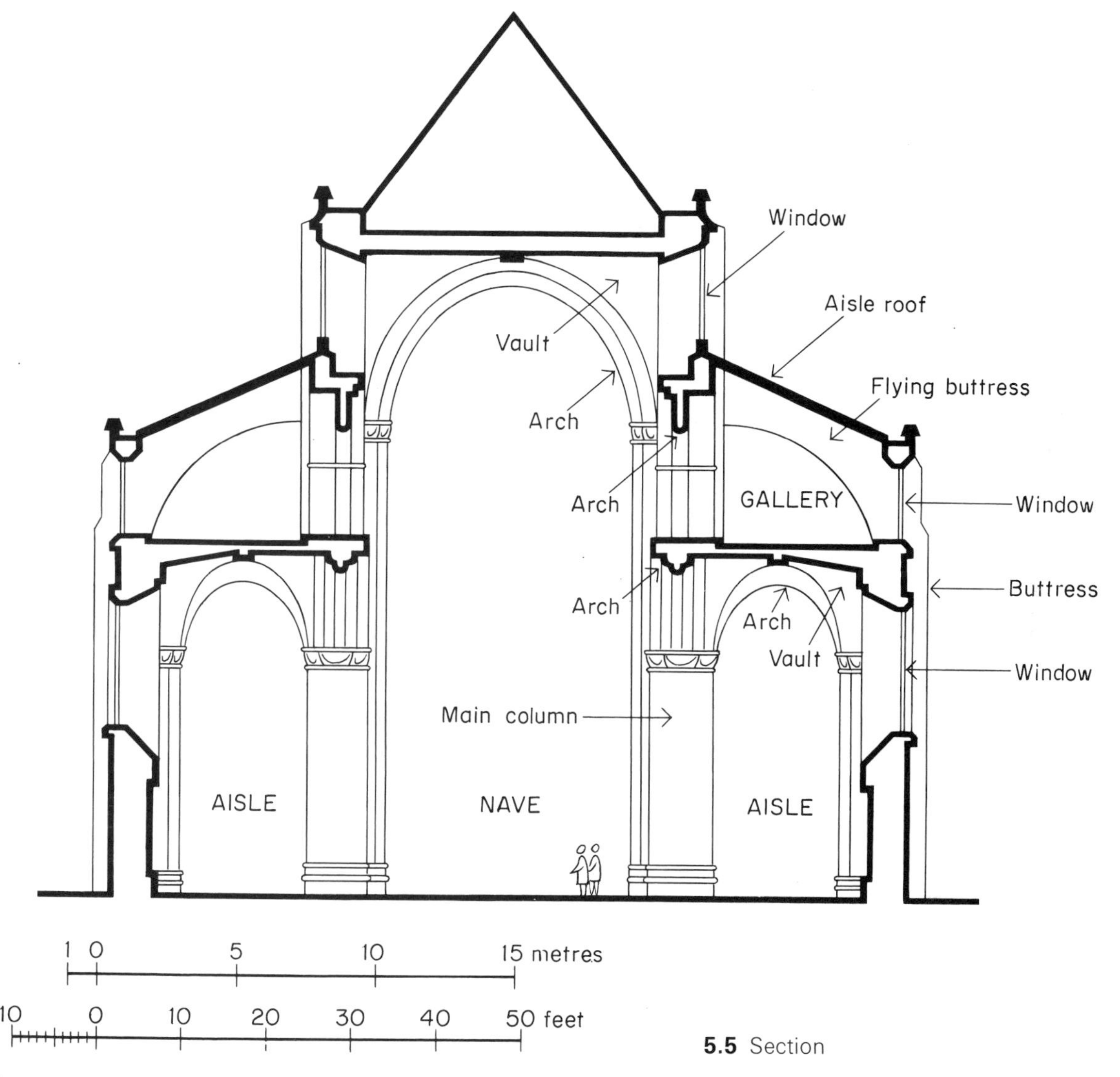

5.5 Section

it also caused trouble. The half-circle rib over the short side of the rectangular space below had to be propped up on stone **stilts**. These stilts brought the rib up to the same height as the rib over the rectangle's long side. If this stilting was not done the two ribs had different heights due to the difference in their diameters. This made the curves of the vault's infill surfaces difficult to build. But though a stilted vault could be built easily and quickly it looked wrong when finished.

This problem was solved by the use of **pointed** ribs. A pointed rib was easy to build because its curve too was a segment of a circle. The segment was used twice, first for one half of the rib and again for the other half. The two segments cut into each other at the top of the rib giving it the pointed shape. (*Figure 5.7.*)

So a pointed rib or a pointed arch had two segments from two circles with the same radius but different centres. The greater the distance between the two centres, the steeper and more pointed the rib or arch became. But the closer the two centres came to each other, the lower and less pointed the arch became. This led to the most useful fact about pointed ribs. Two ribs of different spans could have the same height. This fact gave designers a new freedom for they could put vaults over any rectangular space, and so the whole building could be freed from the rigid grid of squares needed for Classical vaults.

Europe's first pointed ribs for a vault were built at Durham Cathedral in 1128. They spanned more than 9·75 m (32 ft) at right angles across the nave. They were more than 21·4 m (70·3 ft) high from floor to underside of the rib. The other ribs in the vault were the usual Norman type made from one segment of a circle.

Durham's few pointed ribs did not make the cathedral's style Gothic rather than Norman. The Gothic style was later to be a whole system of pointed

Façade

South Aisle

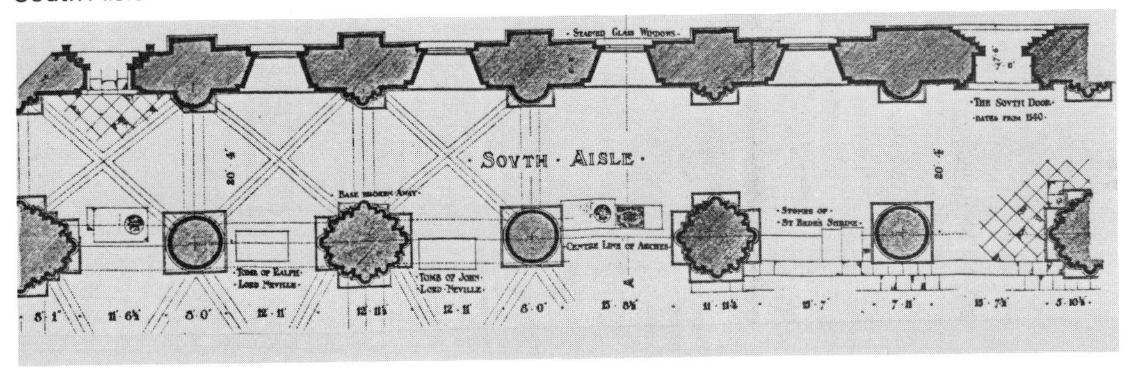

5.6 An inside part of the three-storey nave wall and vault

arches and ribs. The Normans did not invent the pointed arch either. It had been used in 722 BC over a drain at Khorsabad in Iran and of course the Normans did know that pointed arches had been a main feature of Islamic buildings.

In no way did the Norman style at Durham make the cathedral look like a Roman building. Yet by using axis lines and grids of rectangles the Normans showed that the ideas of the Classical style had been kept alive in Europe if not in Britain and, like the Classical designers before them, Durham's builders were also concerned about outlines. The arches and the vault ribs were rounded with convex and concave surfaces and decorated with three-dimensional patterns. The huge round piers in the nave had bold zigzag and other patterns cut deeply into their surfaces.

Where two different parts of the construction came into contact the Normans placed a special feature. For example, they put a large block of stone where the nave arches and the tops of the columns joined. The block was a **cushion** capital and its shape changed from square on top to round underneath. It was used as the Classical designers had used their column capitals. The Normans' cushion capitals not only gave firm support to the arches but also showed that the joint between arch and column was an important part of the structure. The same idea was used where the vault ribs rested on the walls.

When a structure's main points of contact were shown up, as on Durham's cushion capitals, then the structure was **articulated**. Joints and changes from one material to another could be articulated. So could the change from the structural use of a material to a non-structural use of it, such as the Durham vaults where the ribs were decorated but the infill surface was plain. The articulation between rib and infill was to become one of the main features of Gothic vaults.

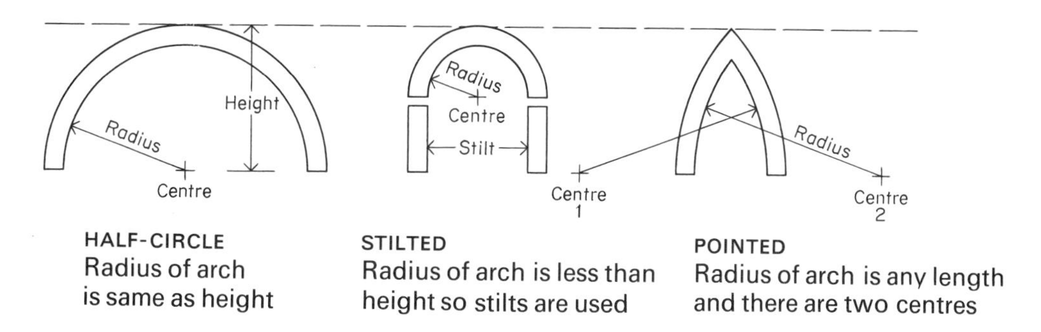

5.7 The basic geometry of half-circle arches and pointed arches

Durham did not need articulation to be stable and strong, but it did need articulation to make it *look* stable and strong.

Work on the site at Durham began in 1093 at the east end and went towards the west end. This was the method Normans used for large churches. The whole eastern end of Durham's hall was built first and it took 11 years to complete. The shorter hall was almost finished by 1099 and its vault by 1133. The massive piers and arches were also ready in 1099 to carry the central tower. The rest of the longer nave west of the central tower took another 32 years in several stages.

In 1133 the cathedral could be fully used. It had taken 40 years to build, but there was still more to come. Between 1175 and 1190 a large porch was built on the cliff edge at the site's west end. By 1226 the two west towers had been built up higher though the tops on them seen today were not added until 1780. They and the central tower were meant to have spires but never got them. In 1429 lightning struck the central tower and it took twenty years to rebuild it 61 m (218 ft) high between 1470 and 1490. Alterations and additions to the other buildings on the cathedral's south side went on until 1539 when King Henry VIII took control from the priests.

A large building site today is planned by critical path networks and computers. It is a busy scene of cranes and concrete batching plant. Earth-moving machines and mechanical diggers make ready for the foundations. Huge trucks come and go with steel beams and tonnes of cement. Power tools whine and electric lamps glare into the night. But Durham's vast project only had man-power.

It is not known how long it took to dig out the foundations by hand. Some of the east wall's foundations went down 4·3 m (14 ft) before solid rock was reached. There was bad ground at that eastern end of the site. But the Normans knew how to deal with it.

Sites for cathedrals were not chosen because they were sound but because they were sacred. Many of these large churches were built on a site where a much smaller church had stood before, or on sites which had links with rites going back to Stonehenge and beyond. Many of the cathedrals were built on wet ground near rivers on which boats could bring stone and wood for big building schemes.

The Normans began Winchester Cathedral in 1079 and Ely Cathedral before 1090. Both these buildings were on bad ground. The designers knew the nave piers would bring the weight of massive masonry down, so on top of the foundations they put a network of low walls called **sleeper** walls. The sleeper walls helped to spread the heavy load from each pier over more of the foundations and, because the loads were spread, the weight on any one area of poor ground was less.

But there was another method the Normans used for making foundations stronger on bad ground. This was the method of **bond timbers**. Baulks of wood at least 305 mm (12 in.) square were framed together by carpenters and laid flat in the stonework below floor level and completely covered in by masonry.

The Romans were using bond timbers a thousand years before the Normans. An example was the bond timbers in the foundations of Pevensey Castle built on marshy ground near the sea in Sussex; and before AD 900 the Anglo Saxons put up a church at York with bond timbers in its foundations.

Norman designers used ideas from the Classical style such as axis lines and grids of rectangles and outlines. But the Normans did not know about the Roman method of building with concrete and, as a result, made their own thick walls like a stone sandwich with rubble filling. Two walls of squared stone blocks set with mortar in level courses were filled in with rubble or flints set in mortar.

When well built, such walls stood for hundreds of years and many still stand today. But some Norman walls were not good. At Durham the vaults at the east end of the building began to show bad cracks only a hundred years after they went up. So in the early 1200s the east end vaults were taken down and most of the Norman work there removed. In spite of the bad ground a new and second hall was built at right-angles to the cathedral's main axis line. This extension was finished by 1280. All its arches and vault ribs were pointed because by that time the Norman style had given way to the Gothic style.

6 Salisbury Cathedral

Place Salisbury, Hampshire
Time *Start* 1220
Finish 1266
Construction time 46 years
Purpose A church. But also like Durham, a cathedral
Style Gothic. Early English

The man who designed the Gothic extension at Durham had already designed Salisbury Cathedral. He was Elias of Dereham who, along with Nicholas of Ely and many other masons, built Salisbury Cathedral in the graceful Gothic style called **Early English**. (*Figures 6.1 and 6.2.*)
Large Gothic churches had the same basic layout as Norman ones. The rib-vaulted naves crossed at right angles and the crossing carried a tower. (*Figures 6.1, 6.3 and 6.4.*) The nave walls had three storeys with rows of arches and piers that took the vault's weight to the ground. (*Figure 6.5.*) Buttress walls at right-angles to the building took the vault's outward push. (*Figure 6.6.*) In the Gothic style, however, all the arches and ribs were pointed and the vault supports were **flying buttresses**. These mid-air stone bridges took the vault's thrust over the aisle roofs and onto the vertical buttresses. Flying buttresses never became as vital to design in England as they did in France where designers made these mid-air bridges into skeletons of stone around the outside of cathedrals. At Notre Dame in Paris rows of high bridges spanned onto massive vertical buttresses. (*Figure 6.7.*)

The vertical part of the flying buttress often carried a counterweight on top which was built as a small solid spire called a **pinnacle**. Many Gothic cathedrals gave the spiky effect to their skylines from rows of pinnacles. Salisbury's pinnacles were kept to the tower and the gable ends of the roofs.

In England designers worked in the Gothic style during two main periods.

Period 1 – 1200 to 1500 in three phases known as Early English, Decorated and Perpendicular.

Early English: 1200 to 1300 Salisbury Cathedral was built in this style where the vault ribs and arches made a design with outlines and surfaces rather than with the massive three-dimensional volumes of Norman design.

6.1 Looking west down the three-storey nave. All the arches and vault ribs are pointed in the graceful Early English style. Slender shafts cluster around the main piers. The dark pier on the left is a tower support

6.2 Looking at the south side and the Chapter House
6.3 The Chapter House. The large windows with tracery show the building to be in the Decorated style

Deeply undercut mouldings gave the curving outlines and surfaces clear articulation. The top storey of the nave walls, known as the **clerestory**, had tall and narrow **lancet** windows in groups of three. Salisbury's three storeys were 12·8 m, 4·8 m and 7 m high (42 ft, 16 ft, 23 ft).

Decorated: 1300 to 1400 Salisbury's Chapter House was built in this style. Most cathedrals had a chapter house close to the church as a free-standing building for priests to use as a meeting room. The chapter house at Salisbury was an octagon with large windows and a vault supported by a slim central pier.

The Decorated style had many more outlines and surfaces of arch and rib than Early English designs. Clerestory window areas were enlarged by taking space from the second storey of the nave known as the **gallery** or **triforium** which then became lower. The large clerestory areas of glass were stiffened against wind pressure by stone frames called **tracery** built in patterns of interlocking circles and pointed arches. In the Decorated style the building's basic construction was used as decoration.

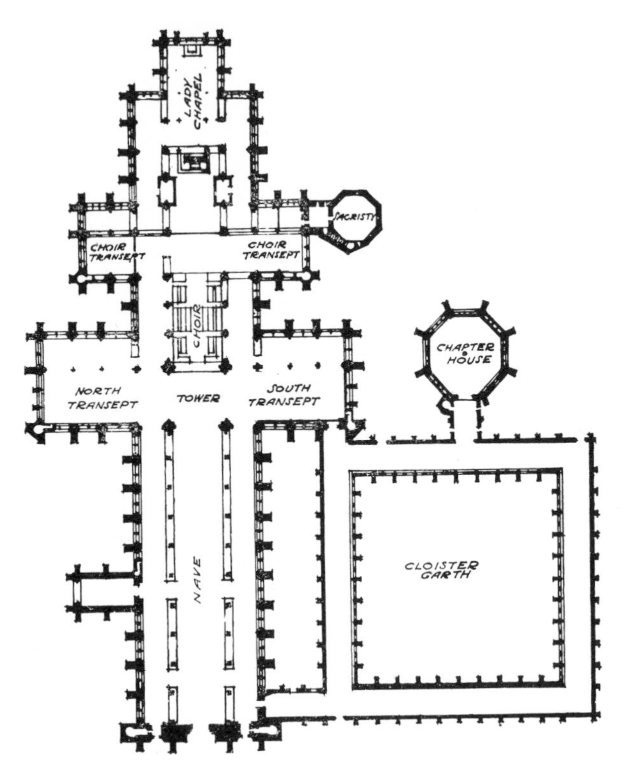

6.4 Plan

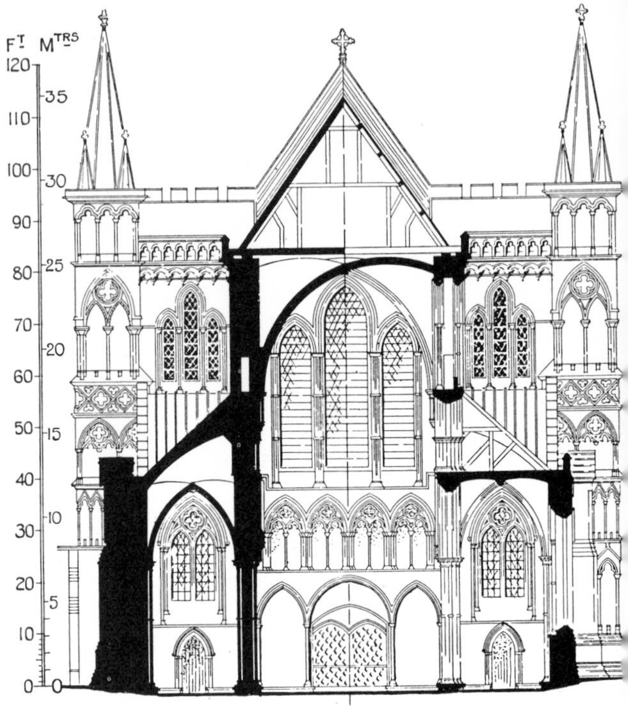

6.6 Half sections

6.5 One bay of nave, inside and outside view. From *History and Antiquities of the Cathedral Church of Salisbury* by John Britton. Plate XIX

6.7 Flying buttresses at Notre Dame in Paris

6.8 Looking up one of the piers which supports the central tower. The pier has bent under the tower's weight

Perpendicular: 1400 to 1500 King's College Chapel at Cambridge was designed in this style which made the inside of churches look as if they were nothing but a huge vault held up by high piers. There seemed to be no walls but only vast areas of stained glass set in tracery between the piers. Everywhere the outlines and surfaces appeared vertical or **perpendicular** and went from floor to vault without a break so that the vault ribs picked up the outlines and took them over the space in sweeping curves. (*Figure 7.1.*)

The inside space seemed weightless but rows of heavy buttresses outside took the vault's thrust down to the ground. (*Figure 7.3.*) The windows had become so wide that the stone tracery had to be a complete wind-resisting frame. This again brought out the perpendicular outlines. In the Perpendicular style the walls looked like non-load-bearing panels rather than load-bearing walls. Every surface of wall inside and out had patterns of perpendicular panels carved in the stone.

From 1200 to 1500 Gothic designers worked towards this system where the vault seen from inside appeared to float on piers too slender for the vault's weight. But this striving for structural slenderness did not reach perfection until Victorian times when the Palm House of 1844 went up in London's Kew Gardens. This was a building with no walls at all but only sheets of glass covering rows of thin iron columns and thin iron arches.

In the 300 years after 1200 the Gothic masons tried to find ways of using less and less stone not only as a challenge to their skill but also as a way of cutting costs. At 124 m (404 ft) Salisbury had the tallest tower and spire in England yet the piers holding them up were among the thinnest. Canterbury's tower rested on 3·65 m (12 ft) thick piers. York's were 3 m (10 ft) thick and Worcester's 2·75 m (9 ft). Peterborough's tower fell down. But the piers carrying it were too thin, being only 2·13 m (7 ft) thick. They were the same as Salisbury's. (*Figure 6.8.*)

The two upper stages of Salisbury's tower were not begun until a hundred years after the main parts of the cathedral. The original tower only rose just above the roof ridges so the piers at the crossing may never have been intended to carry the weight of England's tallest spire. Yet in 1330 masons began the thin shell walls of the tower and the even thinner stonework of the spire. The spire's sloping sides were only 229 mm (9 in.) thick but a wooden framework inside stiffened them against wind pressure.

In the 1200s masons like those working at Salisbury did not calculate thicknesses and sizes in the same way as today's structural engineers. Many Gothic structures fell down. Yet masons must have had methods of some sort but, as they kept them secret, it is not known now what they were. It is thought their design methods may have been geometric rather than arithmetic.

Period 2 – 1750 to 1950, often called the **Gothic Revival**.

This second period of the Gothic style began and ended with little activity but reached a peak of Gothic design and building from 1830s to the 1890s. The Houses of Parliament at Westminster designed in 1835 was the first important public building in Gothic Revival style. The designers Charles Barry and Augustus Pugin borrowed the Perpendicular style from King Henry VII's Chapel just across the road at Westminster Abbey. (*Figures 9.1 and 9.2.*)

At first the Gothic Revival was no more than a style of decoration and ornament. By the 1850s the style was seen as a serious method of construction for all types of buildings. After 1900 nobody used it much at all except for churches and by 1950 following the Second World War the Gothic style came to an end.

Gothic Revival designers learnt how to use the Early English, the Decorated and the Perpendicular styles as well as French, Italian, German, Belgian, Dutch and Spanish Gothic. They went even further back to the Norman and Romanesque styles and kept on going until there were no more styles from the past left for them to copy.

7 King's College Chapel

Place Cambridge
Time *Start* 1446
Finish 1515
Construction time 69 years. But work stopped for long periods.
Purpose A church. When a church is part of a college it is called a 'chapel'
Style Gothic. Perpendicular

King's College Chapel is one huge vaulted room measuring 88 m (289 ft) long and 14 m (45 ft) wide and 24 m (79 ft) high. (*Figure 7.1.*) It is one of the world's most beautiful man-made spaces and is famous also for its music. People who go in to see the chapel, instinctively look up at the great vault covering the whole space. It is breathtaking. (*Figures 7.1, 7.2 and 7.3.*)

Although built 400 years after Durham, King's vault spanned only 3 m (10 ft) more than Durham's. A vault's span is limited by its weight which increases rapidly for quite small increases in span. As a vault becomes heavier its outward thrust becomes greater and so needs bigger buttresses. A span of 14 m (45 ft) seems to have been about the technical and economical limit. The vault at Beauvais in France also spanned nearly 14 m (45 ft) but it needed flying buttresses with three bridges each in the 48 m (157 ft) height, and even then the vault fell down in 1284 hardly 60 years after the cathedral was begun.

In Spain the cathedral at Gerona was built in 1458 with Europe's biggest Gothic vault. It spanned 22·5 m (73 ft) but needed a 6 m (20 ft) thick wall to resist its massive outward thrust. Although King's College Chapel did not compete in height or width with its French and Spanish rivals it was more practical and more beautiful than either Beauvais or Gerona. Its structure is still seen as daring today. (*Figures 7.4 and 7.5.*)

King's College Chapel is one of the last great Gothic vaults in England. The difference in design between Durham Cathedral and the chapel is startling. Yet by 1500 the method of building Perpendicular vaults had almost gone back to the methods of barrel vaulting used by the Romans 1400 years before. (*Figure 7.6.*)

During the Gothic style's Decorated phase from 1300 to 1400 more and more ribs had been added to vaults until there was hardly room left for any more on

7.1 Looking east inside the single great space

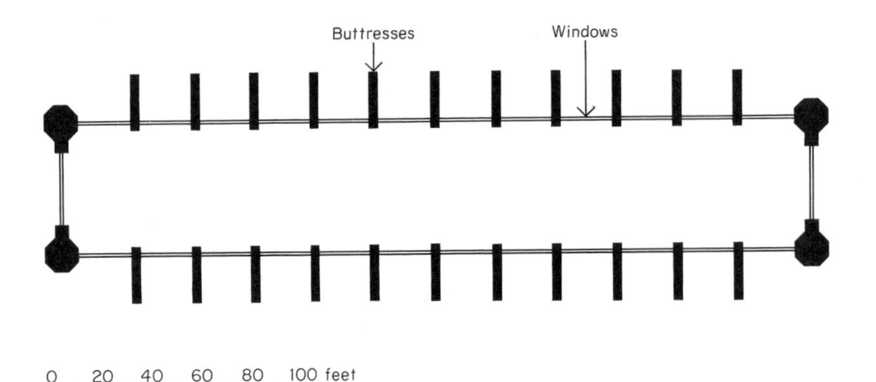

7.2 Plan

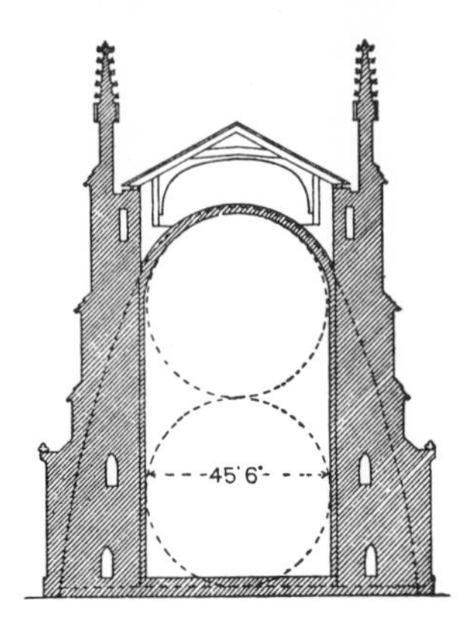

7.3 Section

the vault's curving surfaces. This led to the beautiful design called a **fan** vault. The basic unit of a fan vault is an upside-down half cone of stone. The half cone is concave and is made up of vault ribs which all have the same curve. Cones on opposite walls touch in the centre of the vault.

At King's College some of the cones were built up with separate ribs but others were built like barrel vaults as an unbroken eggshell of stone. The ribs of these cones were carved to give surface decoration. Another famous fan vault covered King Henry VII's Chapel at the east end of Westminster Abbey which was built between 1502 and 1512. The method was that of the barrel vault, yet when seen from below the vault is a lacework of fake ribs carved on the barrel vault's surface.

The first fan vault of all was built at Gloucester Cathedral in 1377 though it only spanned 3·65 m (12 ft) over the walkway known as a **cloister**. The first fan vault over a wide span was at Sherborne in 1475. In 1508 the fan vaults at King's College Chapel were begun. When finished in 1515 they had become one of the world's wonders.

Richard of Ely made the first designs for the chapel around 1445. This master mason-designer used ideas he had seen at Ely Cathedral where a small chapel had been built between 1321 and 1349 and which was like a scale model of the vast chapel King's College would build later. The stone piers supporting the vault over the little chapel at Ely were as thin as their master mason dared to make them.

Richard of Ely took these ideas to Cambridge where all the foundations were laid in one sequence of work, unlike most Gothic churches which went from east to west in stages. A hard kind of chalk called **clunch** went into the ground as the chapel's foundations.

John Woolrich became the master mason in 1476 though after 1485 work stopped for 20 years as there was no money. It began again under John Wastell of Bury St Edmunds who had already been the master-designer of Canterbury Cathedral's central tower. He started on the site at King's College in 1508 and worked out the vault's final design and construction and the method of getting the huge stone cones up into position.

As usual with large Gothic churches King's chapel walls began at the east end and went towards the west end. This method can still be seen at the chapel today where a change in stone colour goes in steps downwards from east to west. At the west end overlooking the River Cam the stonework is lighter in colour. At the east end most of the stone is darker. This change in stone colour shows where the building work stopped for 20 years.

The two kinds of stone came from two different quarries. The chapel's east end was mainly built with a white magnesian limestone from Tadcaster in Yorkshire and the west end with limestone from King's Cliffe in Northamptonshire.

Transport by water was the easiest way for stone to go from quarry to site. Transport costs were high. There were no easy routes by road for many overland tracks could only be used for a few weeks in summer. So any heavy loads went by river and sea just like the bluestones went from Wales to Stonehenge over 3000 years before.

Stones for building could go from quarry to site more easily if they were small rather than large. That is one reason why the Norman builders of Durham used stones measuring about 300 mm (12 in.) on each face. In the 1440s Eton College bought stone from Huddlestone in Yorkshire for its new chapel by the River Thames and paid more than six times for the transport than for the stone itself.

The grids of tall rectangles which helped to give the Perpendicular style its name made a unity of surface design at King's College Chapel and extended over walls and windows and vault. (*Figure 7.7.*) The wall rectangles were carved with designs of interlocking pointed arches and they served as a frame for sculptured

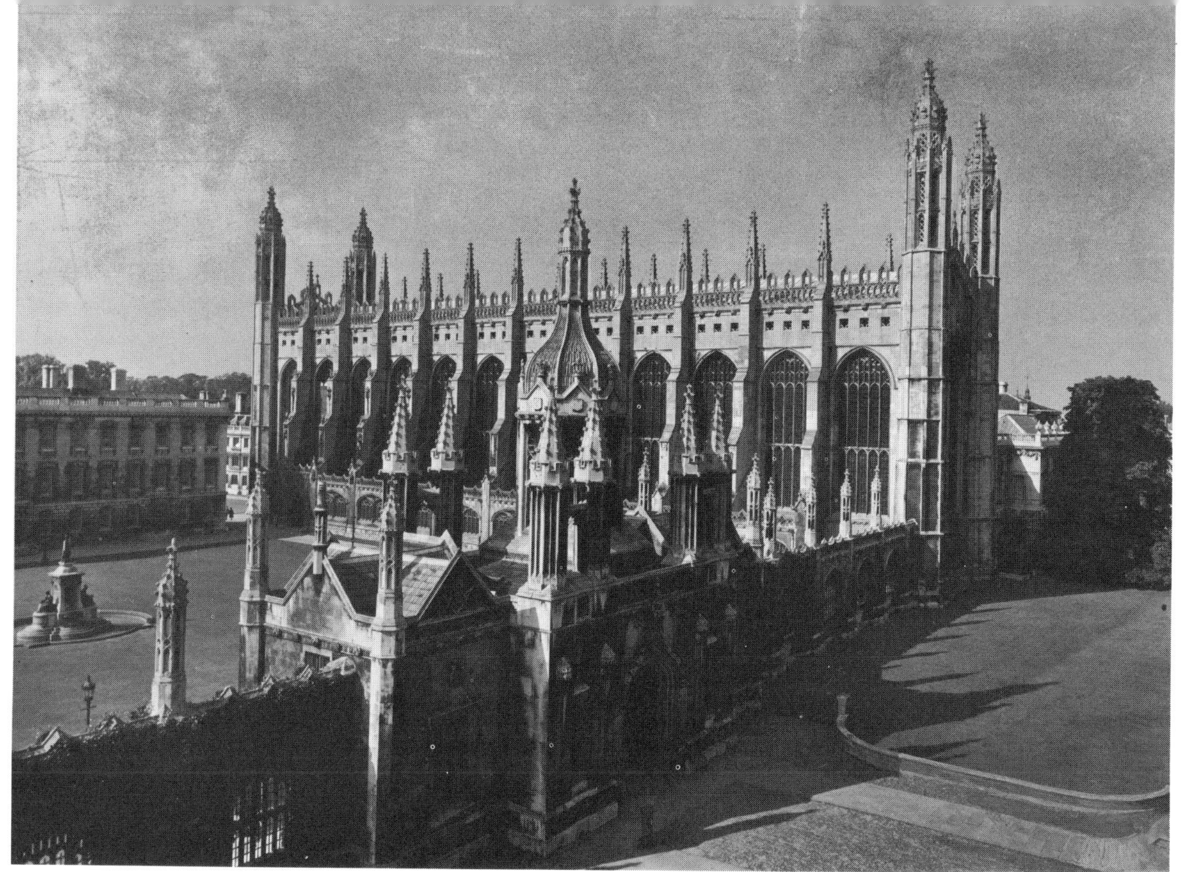

7.4 The windows, buttresses and pinnacles of the south side. There is almost no wall

7.5 The west end seen from the river bank. The Classical style building on the right was designed for King's College in 1723 by James Gibbs

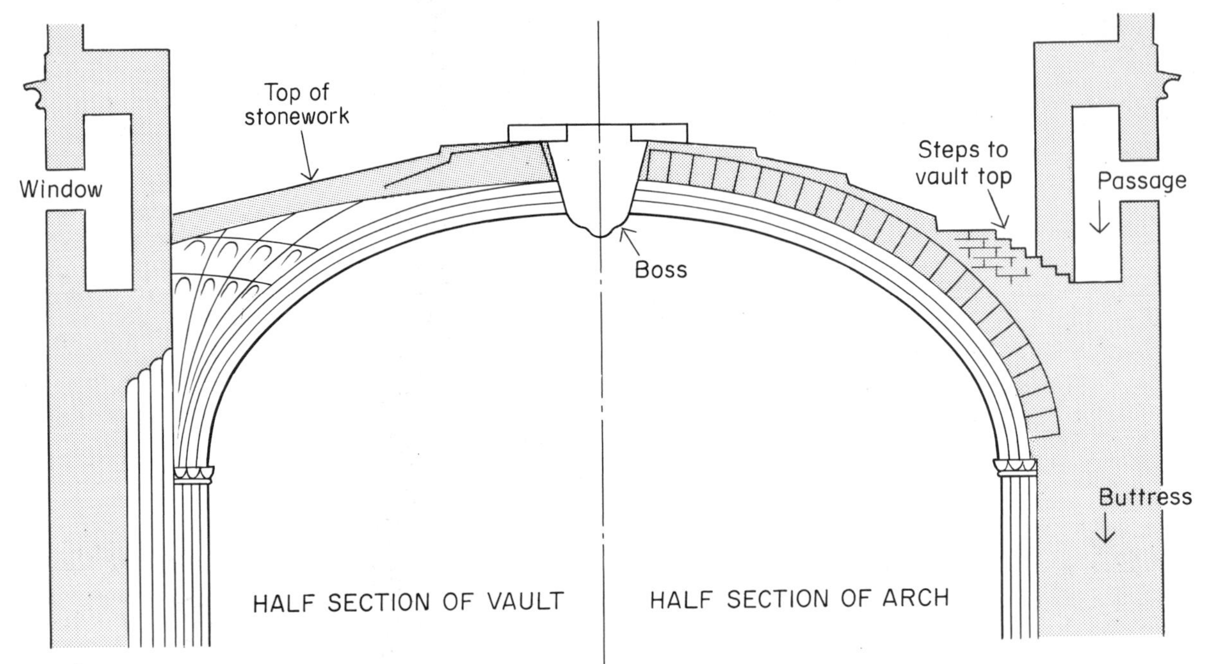

7.6 Detail of vaulting
7.7 An inside part of the single-storey wall and vault

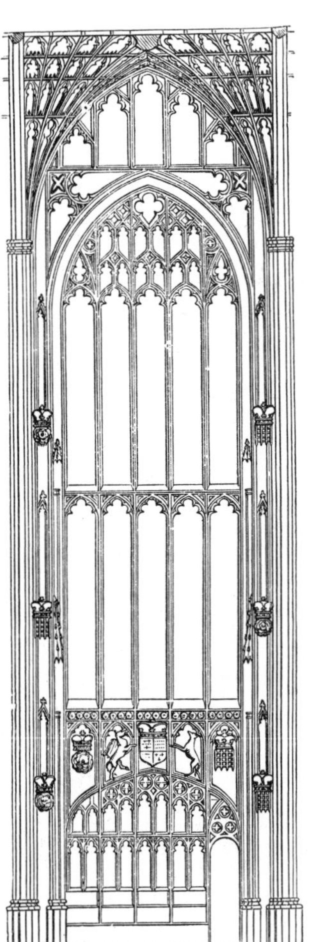

lions and antelopes, greyhounds and dragons, shields, crowns, Tudor roses and other heraldic emblems of the English kings who began and finished the chapel.

The window rectangles made the shapes of the tracery which held the 1206 sq m (13,000 sq ft) of stained glass designed as a hundred vast pictures. The glass was cast in small pieces: the colour fired on one side, the other side left plain. On their benches the craftsmen assembled the pieces into their design and fixed them together with thin strips of lead called **cames** which they bent to follow shapes in the window's picture, such as the outline of a head. The craftsmen then made up the coloured glass and lead into rectangles ready for fixing into the rectangles of the window tracery where thin iron rods helped to hold the glass in place and gave extra stiffness against wind pressure.

The vault rectangles followed the curving surfaces and showed up clearly the solid geometry of the cones. Yet the vault rectangles also seemed to be full of voids, like the tracery rectangles in the windows, and this helped to give the effect of the vault floating weightless in space.

But what the vault looked like and what it was really like were two different things. The weightless-looking vault in fact weighed 1800 tonnes and its outward push had to be resisted by two rows each of 11 outside buttresses turned at right-angles along each side of the chapel. The piers inside which looked too slender even to hold up such an airborne-looking vault were not piers at all but only the inner edge of these mighty stone buttresses carved to look like piers.

To resist the vault's overturning thrust at the top the buttresses had to be 7·6 m (25 ft) wide at the bottom. Low rooms were fitted into this outside space between one buttress and the next and, above their roofs, the buttresses stepped back until they ended in the sky as pinnacles. A turret at each corner of the chapel gave extra stability and drew people's gaze even higher into the sky.

If perfect buildings exist then King's College Chapel is one of them. It was certainly a perfect example of the Perpendicular style. The chapel made stonework weighing thousands of tonnes look as if it floated in air and it made the massive supports which took the weight downward carry the eye upward. Mason-designer John Wastell made the most of this startling contrast between the facts of construction and the effects of construction. (*Figure 7.6.*)

Another designer was to do so with another Perpendicular building. This was Charles Barry with his Gothic Revival design of 1835 for the new Houses of Parliament at Westminster. The effect of the building's Perpendicular panels and pinnacles and the soaring perpendicular lines of its two towers hid the fact that much of the construction was not of stone but of iron. Perhaps John Wastell would have used hidden iron too if he had been starting a building in 1835 instead of finishing one in 1515.

7.8 The fan vaults

8 Westminster Hall

Place Westminster, London
Time *Start* 1394
Finish 1401
Construction time 7 years
Purpose A hall for state occasions and for Parliament. Also used as law courts in the past
Style Gothic. Perpendicular. Hammer beam roof

8.1 Looking down the Hall to the Perpendicular style window at the end

The original walls of this royal hall – the largest in Europe – were built by the Normans. As with some Norman cathedrals, the original Norman walls were built up higher and refaced with stone in the Perpendicular style, and the small Norman windows down each side of the 73 m (239 ft) long hall were replaced by Perpendicular windows with tracery. (*Figure 8.1.*) Master Henry Yevele designed the stonework and was the master mason in charge of it. (*Figures 8.2 and 9.3.*)

Heavy new buttresses of stone were built outside to take the thrust of the huge oak roof covering nearly 0·2 hectares (0·5 acres) of floor area and spanning the whole 20·7 m (68 ft) width of the Great Hall from wall to wall with no support in between. 660 tonnes of oak went into the roof and each point of support on the new walls carried 13 tonnes. Master Hugh Herland was the designer and also the master craftsman of northern Europe's largest wooden roof. Only England had roofs like this and only Westminster had such a masterpiece. (*Figure 8.2.*)

The walls inside had to be made ready to take the ends of the 13 oak trusses. Project planning began in 1393. Stone was ordered from the quarry at Marre near Doncaster in Yorkshire and two masons, Swallow and Wasbourne from Gloucestershire, were given the contracts for building the stone ledge halfway up the walls and which went down each long side of the Great Hall just under the new windows. The two Gloucestershire masons also built Marrestone corbels into the ledge from which the huge trusses began their upward curves. Later Swallow joined Yford, another mason, and together they built at least 12 of the 24 new windows.

The oak trusses rested on the wall corbels every 5·9 m (19·5 ft). At Westminster Great Hall each truss had three main parts. (*Figure 8.3.*)

1 Main rafter and collar beam

The bottom ends of the rafters rested on the walls and the top ends were joined together. To stop it from spreading outwards each truss was tied halfway up by an oak beam measuring 762 mm × 584 mm (30 in. × 23 in.). This was the collar beam.

2 Main arch

This was a huge pointed arch like a stone vault rib rising up in mid-air except that no other vault in Britain spanned 21 m (69 ft) as did this arch of oak which sprang from the stone ledge halfway up the wall. The point of the arch joined the middle of the collar beam.

3 Hammer beams

The truss got its name from these short beams which projected horizontally from the wall. One end rested on top of the wall and the other, sticking out in space, was propped from below. That end carried an upright strut which went up to be framed into the joint where the rafters were tied by the collar beam. The mid-air end of the hammer beam also carried the bottom of a second pointed arch. This was another heavy oak rib like the main arch. Its two circle segments rose up from the end of the hammer beam and joined the point of the main arch.

The curving parts of each truss made the shape of interlocking pointed arches which was a common

8.2 Plan and section

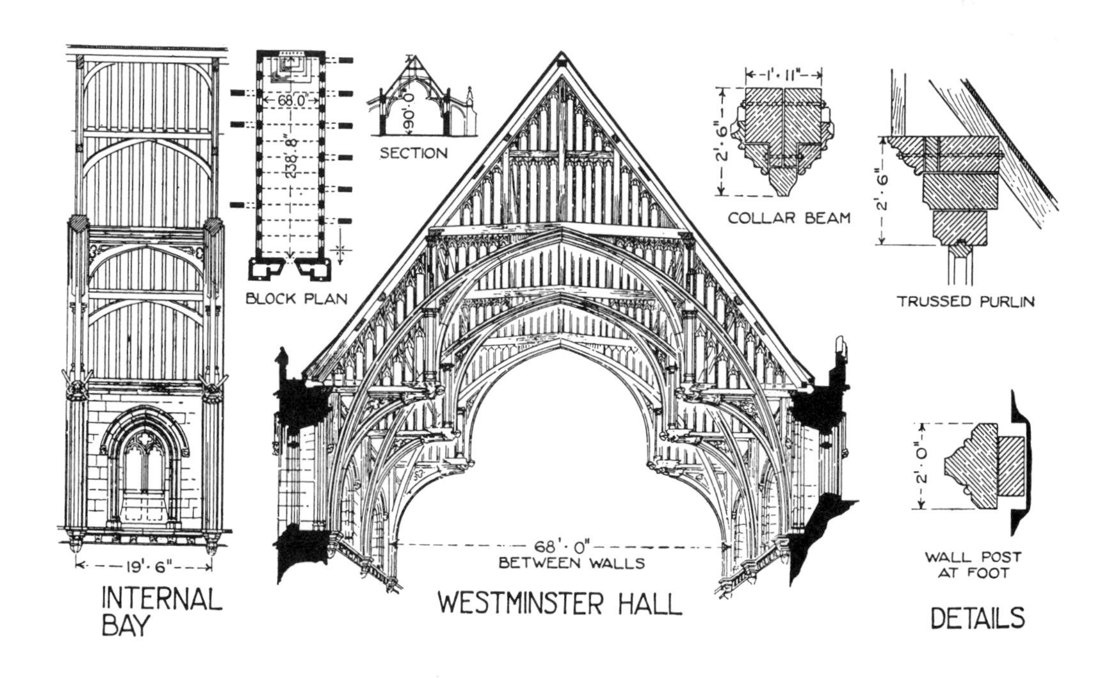

feature of Gothic design. The shape was also used in the tracery of the Great Hall's new windows.

While the masons went ahead with the walls, Master Hugh Herland was hard at work getting the great roof ready. But not at Westminster. His carpenter's workshop had been set up out of London altogether and was at Farnham in Surrey. When he designed the roof Hugh Herland also had to plan how to get it from Farnham to Westminster. Each truss weighed over 50 tonnes.

8.3 The roof's main parts

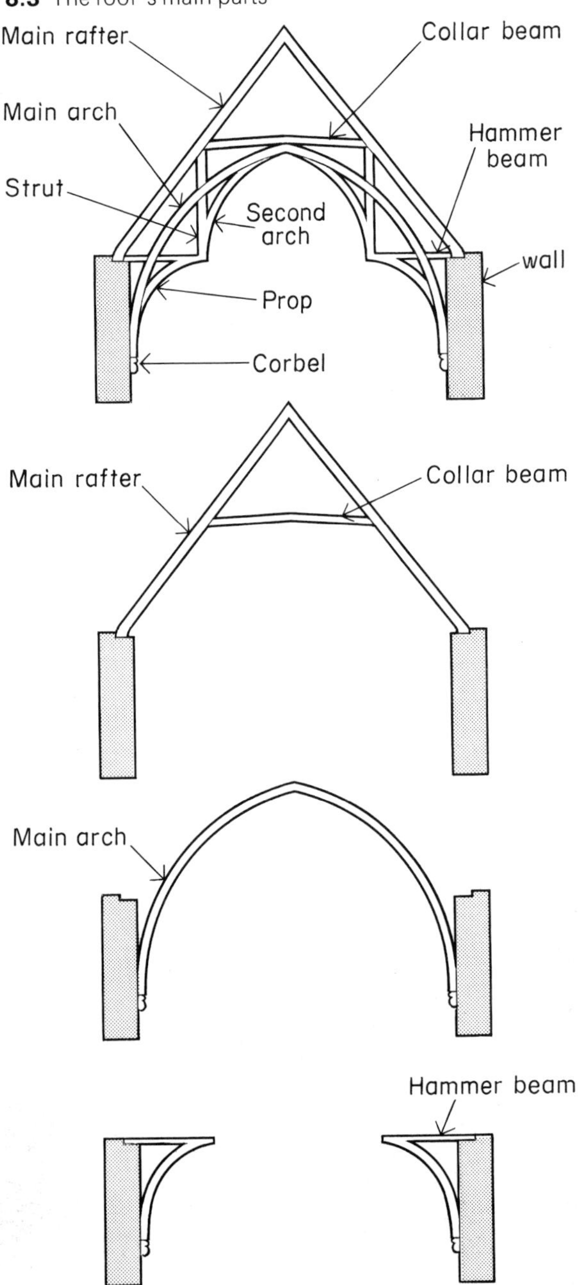

Herland knew that the massive trusses would have to be assembled on site at Westminster from numerous smaller oak parts, and the parts would have to be small enough to go by cart from Farnham to the River Thames and then by boat down to Westminster. So he designed his giant kit-of-parts with this in mind and then, with his carpenters, set about making it.

Some of the oak came from the king's forest at Odiham and Aliceholt in Hampshire. Two hundred oaks to make rafters were bought from the priests at St Albans who owned forests at Northam in Hertfordshire. And William Croyser sold oak from his land at Stoke D'Abernon in Surrey.

Sawyers and cutters at Farnham slowly turned the oaks into pieces from which the thousands of truss parts could be made. Most of the sawyers and cutters were paid a daily rate, but special carvings were paid for by the piece. This was called **taskwork**. And that was how Robert Brusyngdon, William Carron, Peter Davyn and Hubert le Villers were paid for carving the flying angels fixed in mid-flight at the end of the hammer beams.

The tools used by Hugh Herland and his carpenters in 1394 were very like those used at Durham Cathedral 300 years before. The oak logs could not be used for carpentry until they were cut into planks and beams. Two sawyers did this work with a pit-saw which had its long blade fixed in a strong but light wooden frame. One man stood over the log and the other man stood in a pit under it. One pushed the saw and the other pulled it. Skilled men kept the blade in tension so that it did not snap. But it was slow work and hardest for the sawyer below in the pit.

Since Roman times smaller baulks of timber had been sliced into planks by a frame-saw but after AD 1100 the saw was improved by a cord tightened around the frame to keep the blade in tension. The blade could twist away from the frame so that it could cut a plank right through from end to end. Sawyers marked lines on both sides of the wood so they could see where to cut. They coated a string with chalk and stretched it along the wood and flicked it to make a chalk line. There were also saws with two handles and one type like a large breadknife which cut on the down-stroke as hand-saws do today. But sawing with them was harder work in those days as there was no grip handle, only a round one.

Hugh Herland's carpenters did much of the shaping work with their axes, a type of tool not seen today yet already thousands of years old when Roman carpenters shaped the Sussex oaks for the foundation piles and roof trusses at Fishbourne. For the Great Hall at Westminster the carpenters trimmed and squared the oak planks with special trimming axes.

An axe's cutting edge was crescent shaped and had

8.4 The entrance from the Hall into the Houses of Parliament designed by Charles Barry in 1835

pointed ends like a battle-axe and a wooden handle called the **shaft**. Before AD 1300 axe shafts were rectangular in section but by the 1390s Hugh Herland's men had oval-shaped shafts like today's axes, but unlike today's their shafts were straight not curved.

Another tool for shaping wood was the adze. This was like an axe but had a narrower blade which was good for chipping out curved surfaces. Over 4000 years before Westminster Great Hall was built, Egyptian craftsmen had trimmed with the adze.

Fine and smooth shapes were worked by a draw-knife. The carpenter held it by the handles at each end of the blade and drew the knife towards him. The draw-knife was good for curved work such as the rim of a wheel. Planes like today's may have been used for the Great Hall's hammer beam roof. The Romans in Britain worked with planes but the tool was forgotten after they left and it is not known when planes came back again, though it may have been in the AD 1200s. By the 1500s both the large flying plane and the small smoothing plane were in use for fine joinery work. The exact cutting of holes and grooves was done by chisels like today's. Large chisels for rough work had no wooden handles.

Gothic carpenters fixed the parts of large structures together with pegs since nails were not strong enough to hold heavy sections. Holes for the pegs were drilled. Deep holes for joining big pieces of wood together were made by a shell-bit. This half–tube with cutting edges was about 100 mm (4 in.) long and it needed a hole already started by a mallet and gouge. There were also spoon-bits shaped like spoons with round or pointed ends or short screw threads. Spoonbits were no good for deep holes.

Carpenters also drilled with the three tips of a centre-bit. The centre one pierced the wood and so guided the drill and the two outside tips were made as cutting blades. Roman carpenters had these types of bit amongst their wide range of tools and ones like them continued to be used right into the 1800s. Only then was the bit invented which had a spiral groove to throw the drilled wood shavings up and out of the hole.

Deep holes in beams or heavy roof timbers were also drilled with an auger which was a long spindle with a bit at one end and a turning handle at right-angles at the other end. The carpenter could put a great deal of power into each turn of the handle. The bradawl was a small version of the auger and is still used today. Drills also had crank handles with two right-angled bends and by the 1300s the brace with its four bends was invented.

The blades and teeth and cutting edges of all these tools were made of iron by the blacksmith and they had to be sharpened often as iron was not so hard as today's steel tools. Iron edges blunted quickly on hard and close-grained wood like oak so the carpenters sharpened the edges again and again on smoothed rock or slate and went to the blacksmith to have dented and worn edges repaired.

After 18 months Hugh Herland and his carpenters had all the pieces of oak cut and shaped and drilled ready to be taken to Westminster. In June 1395 two carts pulled by 16 horses made 52 journeys from the yards and workshops at Farnham to Hamme on the River Thames near Chertsey. There the oak parts, such as the 26 half beams and the 26 wall posts, were loaded onto boats and went down to the riverside site at Westminster. The carpenters built a big scaffold in 1395 and the first hammer beam truss was assembled and fixed in 1395 and the last one in 1397.

Other wooden rafters ran along the length of the roof joining one hammer beam truss to the next to stop them from falling sideways. So the whole roof became one rigid oak frame. Then the upper surface had to be boarded with oak and covered with lead. In 1401 Westminster Great Hall was finished.

The work had been started in 1394 because of a fire which badly damaged the old Norman hall, and the new Great Hall of 1401 was nearly burnt too in another fire of 1834 which destroyed most other buildings of the Palace of Westminster around it. The old Houses of Parliament were in ruins. In 1835 Charles Barry won the competition to build the new Houses of Parliament on the same riverside site. The Great Hall of 1401 which survived the fire became a vital part of his design. (*Figure 8.4.*)

9 Houses of Parliament

Place Westminster, London
Times *Start* 1837
Finish 1860
Construction time 23 years
Purpose Parliament building
Style Gothic Revival. Late Perpendicular. Also Tudor

9.1 The river front. Victoria Tower on the left. Spire over octagon in the centre. Clock Tower on the right

9.2 The Clock Tower, known as 'Big Ben' was designed by Pugin as was all the Perpendicular detail of the building

Charles Barry designed the Houses of Parliament after the fire of 1834 in such a way that people entered the new building by first having to walk down the whole length of Westminster Great Hall under the hammer beam roof Hugh Herland finished in 1397. Few buildings ever had such a fine entrance as Barry's Parliament and few designers ever made such fine use of an existing building on the site of a new one. (*Figure 8.4.*)

There could be no question about what style the building should have because more than a thousand years of British history in the shape of Westminster Abbey stood at the edge of the site. So Gothic Revival it had to be and Gothic taken from the Late Perpendicular or Tudor style of the 1502 Henry VII's Chapel which was the nearest part of the Abbey. This style for the new Parliament buildings would also blend in with the Perpendicular style of the Great Hall that was a vital part of the scheme with which Barry won the 1835 design competition. (*Figures 9.1 and 9.2.*)

In London Barry had already built the 1829 Traveller's Club in the Classical style and the 1826 Holy Trinity Church in Islington in the Gothic style. But the church was a small and cheap copy of King's College Chapel at Cambridge and its mean-looking details showed that Barry was wise in getting a young designer to do the Perpendicular details of the Parliament competition drawings. And later the same young designer did the detailed design outside and inside and all the rich furniture and fittings from the gilded throne in the House of Lords down to the umbrella stands.

This designer was the 23-year-old Augustus Welby Pugin who at 14 had designed Gothic Revival furniture for Windsor Castle. But Pugin was a rebel and he dressed like a sailor all of the time because he went solo sailing some of the time. He upset his parents and everybody else around him but as nobody else knew as much about design in the Gothic style, Barry simply could not do without him. Pugin was one of the first Gothic Revival designers to see that the main feature of Gothic buildings in the 1200–1500 period was their pointed arches, vaults and flying buttresses. Pugin's work is what people see today in most rooms of the Houses of Parliament except the House of Commons which was bombed in the Second World War and rebuilt to a new design.

To get the site ready for the 275 m × 100 m (900 ft × 330 ft) building the 2·8 hectares (7 acres) were cleared of burnt-out ruins. Along the river front the marshy bank was enclosed by a 3 m (10 ft) thick wall of Aberdeen granite built on 3·6 m (12 ft) deep concrete foundations. The wall ran for over 275 m (900 ft) along the river and turned back at each end to higher and drier land. This giant box of stone was filled to the top with a strip of concrete 8 m (27 ft) wide and 7·6 m (25 ft) deep between the river wall and the foundations of the new buildings.

To keep the water out while work on this embankment went ahead a 281 m (920 ft) long coffer dam was built around it and the first of its 11 m (36 ft) long 304 mm (12 in.) square timber piles was driven into the river bed on 1 September 1837. A 10 hp steam engine running day and night powered two 457 mm (18 in.) diameter pumps to keep the inside of the dam dry. The dam stayed in the river for 12 years and altogether with the top of the platform built up level with the new river wall made a good place for contractors' workshops and for the storing of materials.

Three years later the vast platform called 'The Terrace' was finished and work could begin at the new building's floor level which came to 6 m (20 ft) above the River Thames. By 1841 there were 220 skilled masons on site and three years later there were 700 men and by 1848 more than 1200 were working on the project both at the site and off it.

Charles Barry's overall layout was simple and grand. The great rectangle of buildings was a grid of rooms and courtyards. In the middle of the grid two axis lines crossed each other at right-angles in a 18 m (60 ft) wide octagonal space. (*Figure 9.3.*) Barry knew of at least three other buildings in Britain like that, one Gothic, one Classical and one Gothic Revival.

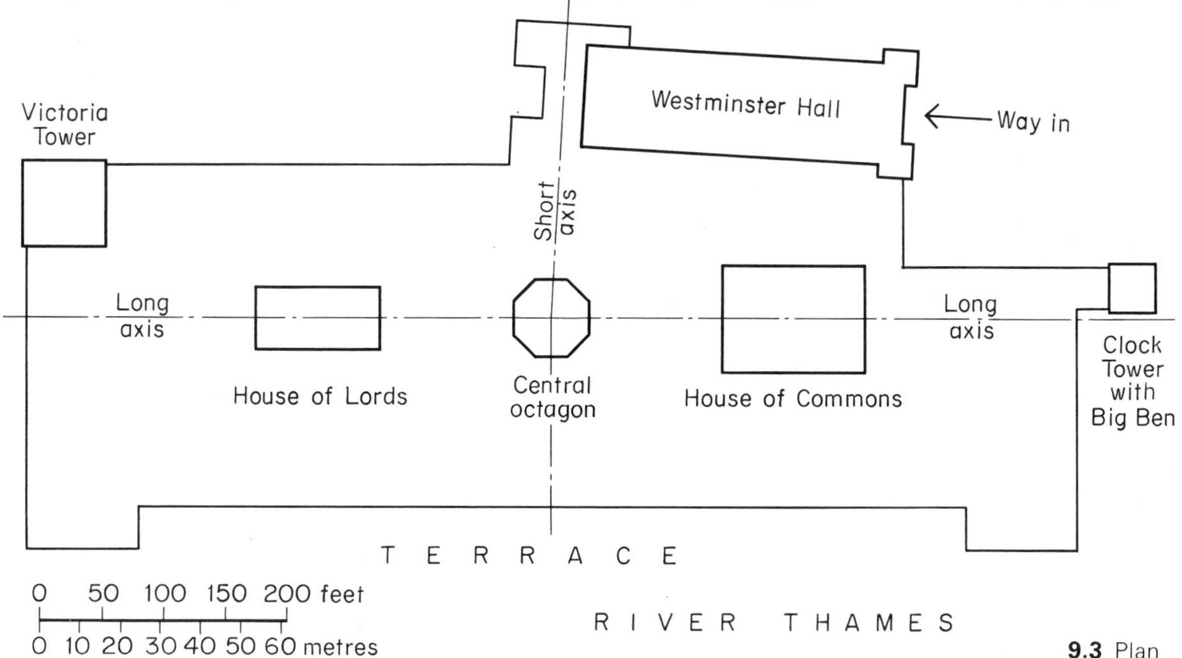

9.3 Plan

1 Gothic

This was Ely Cathedral where the central tower had fallen down in 1321. The ruined piers were cleared away and eight new ones built at the corners of a new octagonal space measuring 23·5 m (77 ft) across. From the top of the piers oak beams leaned inwards to be framed into the bottom of an octagonal tower which they supported. The tower's eight corners were made with vast oak posts 19 m (63 ft) long and 1020 mm × 810 mm (40 in. × 32 in.) in section. The new tower was finished in 1342 with windows in its sides and so it was a **lantern**. (*Figures 9.5 and 9.6.*)

2 Classical

This was the 1675 St Paul's Cathedral in London where its designer Christopher Wren also placed a central octagon at the crossing of the axis lines. (*Figure 12.3.*) Wren knew the Ely octagon well so he too put eight huge piers at the corners of an octagon. At St Paul's it measured 34 m (112 ft) across. But instead of putting oak beams to lean inwards from the top of the piers

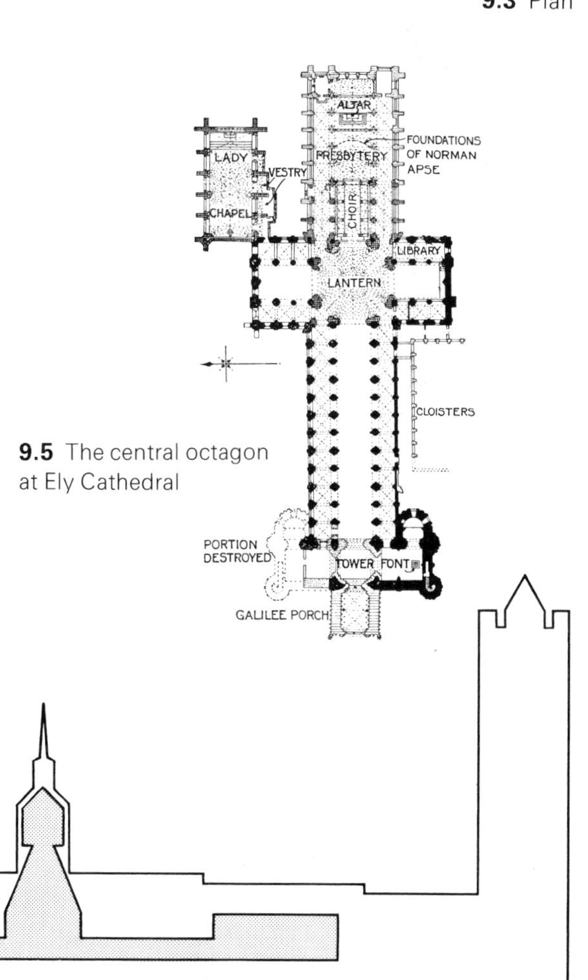

9.5 The central octagon at Ely Cathedral

9.4 Section

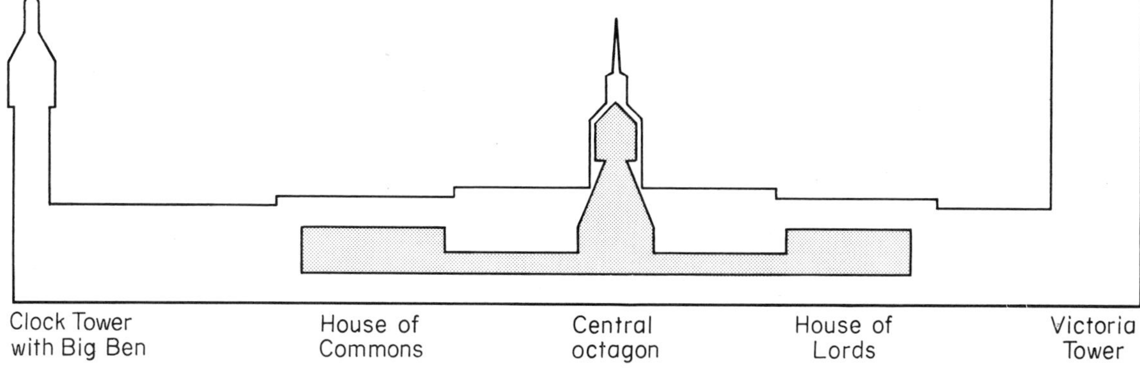

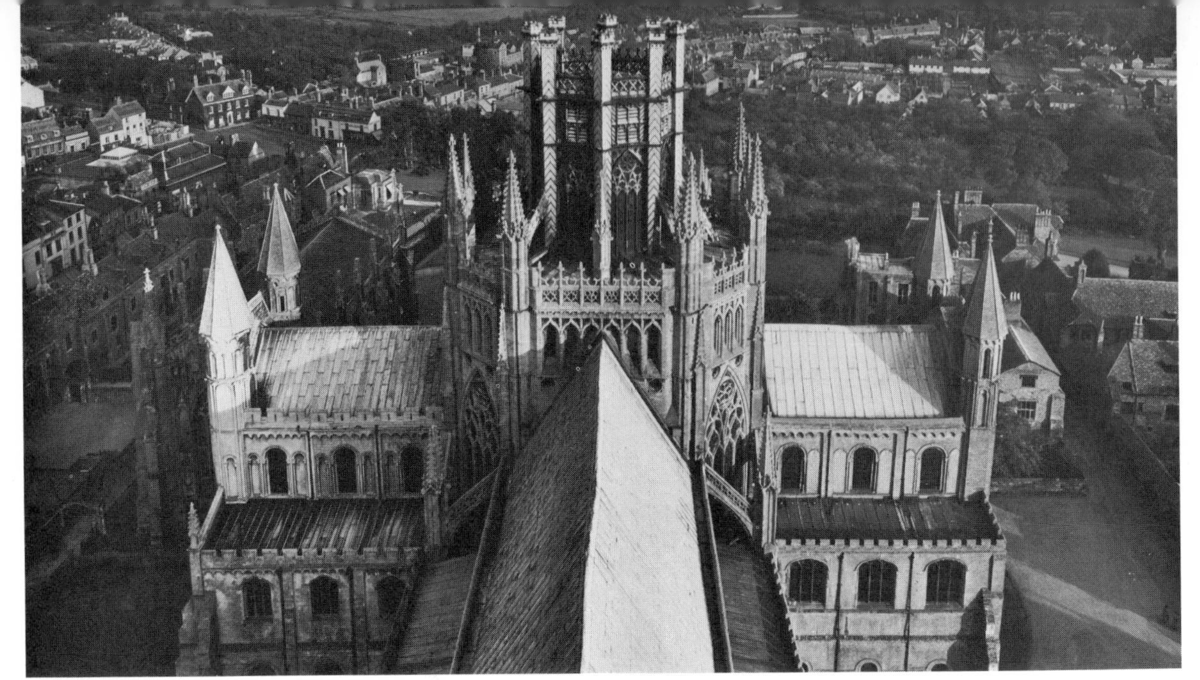

9.6 The octagonal lantern at Ely Cathedral

Wren built the sloping surface of a brickwork cone with its top left off so that it could carry the stone lantern. He may have got the idea for the cone from the 1645 church of the Val de Grace designed by Francois Mansart which Wren saw during his 1666 stay in Paris.

3 Gothic Revival

This central octagon was at Fonthill Abbey, a large new country house in Wiltshire designed in 1796 by James Wyatt. He was called 'Wyatt the Destroyer' because of the ugly way he restored and altered Durham and Salisbury and other great English cathedrals. At Fonthill Abbey he built a 37 m (120 ft) high hall which could not be heated and a 54 m (176 ft) high tower which a storm blew down in 1825.

For his own central octagon at the Houses of Parliament Barry took ideas from the octagons at Ely and St Paul's. Barry's octagonal space was 18·3 m (60 ft) across. Above it he built a 12 m (40 ft) high cone of brick and stone where it carried the stone lantern. Wren had placed a gilded copper ball and cross on the top of St Paul's lantern but on the lantern at the Houses of Parliament Barry put a spire with its top 80 m (261 ft) above floor level. (*Figures 9.1, 9.3, 9.4 and 12.4.*)

At each end of the longer axis line passing through the octagon Barry stood a tower. To the north was the 95 m (310 ft) high Clock Tower housing 'Big Ben' the famous bell which strikes every hour. To the south was the 101 m (331 ft) tall Victoria Tower. This was moved slightly off the axis line to the corner of the building with its fine royal entrance.

On the Big Ben side of the axis line Barry sited the House of Commons with the House of Lords on the Victoria Tower side. Each House has its own cluster of rooms so that they balance each side of the central octagon which is called the 'Central Hall'. People going into the Houses of Parliament through the Westminster Great Hall come to Barry's Central Hall by way of the shorter axis line. (*Figure 8.4.*)

This layout was the first example of a special skill which Gothic Revival designers were to show over the next 60 years. It was the skill to make a simple building from a tangle of rooms. The Houses of Parliament were really several buildings rolled into one. There were the groups of rooms for royal and state occasions and the two Houses, of Lords and Commons, and the two houses for their two Speakers to live in. There were public areas and private areas and office and committee rooms and the library for members of Parliament, and scores of other rooms for running the huge building. Yet Barry was clever enough to put all this in a simple grid of buildings and courtyards with a crossing of axis lines as clear and simple as those at Ely or St Paul's Cathedrals. (*Figure 12.3.*)

Other designers soon followed Barry's example. By the 1870s Londoners could see George Gilbert Scott's station hotel at St Pancras and George Edmund Street's Law Courts in the Strand. Each was a Gothic Revival design and each had a simple layout linking hundreds of rooms. And it was not long before many large towns had built a Gothic Revival town hall and museum, railway station and college and many other types of building that had never even existed in the 1200–1500 Gothic period.

Charles Barry's new Houses of Parliament were also innovative in another way in that of the means of heating and ventilating and lighting by gas. The service systems were an integral part of the building's design and construction and the heated air flowed through duct spaces specially built into the floors and walls and roofs. Even the lantern over the central octagon was not just a lantern but also served as an outlet for the ventilation system.

The Victoria Tower had a turret at each of its four corners and the pinnacles of these were like those across the road in the 1502 Henry VII's Chapel. But unlike the chapel's pinnacles, two of the Victoria Tower's were ventilation outlets. The copper and iron they were made of was not the only metal construction in the Victoria Tower. Inside the 3117 cu m (4270 cu yd) of stonework and 11,402 cu m (15,620 cu yd) of brickwork that went into the Tower's 1·8 m (6 ft) thick walls there were 1300 tonnes of iron. Eight cast iron columns ran through the Tower carrying its nine floors of stone laid on shallow brick vaults between iron joists. The top columns had a 100 mm (4 in.) diameter and the bottom ones 200 mm (8 in.). The walls were built from a climbing scaffold powered by a steam engine and it could also lift 40 tonnes of material at one time. Some stone blocks which masons had been shaping and carving in their workshops for over three months weighed 4 to 5 tonnes. The Victoria Tower was finished in 1860 after 20 years. Charles Barry died in that same year aged 65.

It has since been said that Barry's Victoria Tower was the first skyscraper because of its iron columns and beams and fireproof floors. America's first tall fireproof multi-storey buildings went up in Chicago after the 1871 fire which wiped out the wooden city. Many of the new tall buildings had metal frames inside and masonry outside like the Victoria Tower. But in the true skyscraper the inside metal frame came right through to the outside where only large windows and non-load-bearing panels enclosed the inside. The 1890 Reliance office building in Chicago was built 16 storeys high like that. But even by 1860, the year when the Victoria Tower was finished, there were buildings in America where both the inside and outside construction was a cast iron frame. In 1860 John Kellum built the Stewart shop in New York but as early as 1830 John Haviland built a bank with an iron and glass outside wall at Pottsville.

Pugin died aged 40 in 1852, the year when Queen Victoria opened the new House of Commons which he had equipped with his rich and colourful Gothic Revival decoration and furniture just as he had designed the inside of every other room in the vast building. Pugin too had designed the Big Ben tower, inside which the famous bell had to be lifted up sideways through the Tower's 8·5 m (28 ft) square hollow shaft. Pugin had been one of the first designers who made people see Gothic design as a system of construction and not merely as a source of decoration. But by 1860 iron was taking over as a vital new material for the construction of buildings. The vaults and arches of the Gothic Revival style were limited in span, heavy in weight, slow to build and costly. They could not compete with the wide spans, lighter weight, speed of assembly and cheapness of cast iron columns and beams.

At the Houses of Parliament nobody wanted a repeat of the 1834 fire, so Barry used iron floors and complete roofs of iron with no wood in them at all. The iron trusses were covered by 5 mm (0·18 in.) thick iron plates cast with great accuracy by a foundry in Birmingham and finished on top with a special paint made to a secret formula by a Hungarian called Szerelmy.

But most of the building's iron construction was hidden behind the Gothic Revival stonework. Already there were structures like Gardner's Store put up in Glasgow during 1856 with cast iron and glass outside walls hinting that buildings in the future woud have no stone at all, not in the Gothic Revival style or any other style.

But the end of the Gothic Revival style was slow in coming. Between 1911 and 1913 Cass Gilbert built the 52-storey 240 m (790 ft) Woolworth skyscraper in New York and for 18 years it was the world's tallest office building. But outside it was in the Gothic Revival style with pinnacles and a spire. In the early 1960s the Catholic cathedral at Liverpool was built like the 1342 octagon at Ely with a ring of huge beams leaning inwards to hold up a lantern tower. But unlike Ely the Liverpool beams and lantern were the whole cathedral not just a part of it and they were concrete not oak. The same idea had already been used with curved instead of straight beams at Brasilia's cathedral which Oscar Niemeyer designed for Brazil's new capital city in 1960.

Liverpool not only had its own cast iron office building at Oriel Chambers designed by Peter Ellis in 1864 but also had another cathedral. This was the Anglican one in the Gothic Revival style which was only finished in the early 1980s. It was begun in 1903 by Giles Gilbert Scott, grandson of the man who built the Gothic Revival hotel at St Pancras Station and hundreds of churches all over England.

Gothic Revival designers such as Charles Barry and Augustus Pugin not only left their buildings behind but their ideas too. They thought that design and construction should serve a building's purpose in a clear and simple way and that crossing axis lines and grids of rectangles helped the design process. They agreed that simple solutions to structural problems and the use of any suitable material helped the construction process. They also thought that the Gothic Revival style was the best way to get the best results. No designer today uses their style but every designer today uses their ideas.

10 Little Moreton Hall

Place Cheshire, West Midlands
Time *Start* 1559. *Extended* 1580.
Finish early 1580s
Construction time 1 or 2 years for each stage
Purpose Extensions to a country house
Style Tudor. Also called Early Renaissance

By the 1550s craftsmen were using their skills on houses rather than on churches and as most houses had wooden frames carpenters were kept busy. A master carpenter was usually in control of the whole job. After the wooden frame of a house went up the voids were filled with wicker fencing called **wattle** and then covered with crude plaster called **daub**. Buildings like Little Moreton Hall were called **black-and-white** because of the patterns made by the dark wood and light plaster. Some of the frame's voids were left open as windows, though only rich people could afford glass. Most people just had wooden bars or shutters and hung up oiled linen or animal skins in bad weather. Draughts, dampness and dimness were normal.

In 1559 William Moreton employed carpenters to extend his grandfather's house by adding a gatehouse and rooms with big bay windows, all in the black-and-white style. One of the carpenters carved the inscription: *Rycharde Dale Carpenter Made Thies*

10.1 The courtyard with the wooden bay windows added in 1559

Windovs By The Grac of God. He had tools like those Hugh Herland used at Westminster Great Hall over 150 years earlier.

Little Moreton Hall had no big spans and its frame was made of short pieces of wood fixed close together. There were many small gables and many large windows with wooden frames filled by small pieces of glass in lead cames, making a variety of patterns. By 1590 there were 15 glass works in England and people could choose from a hundred different designs in a catalogue. Richard Dale and his carpenter friends made the whole Hall with black-and-white patterns of stripes and diamonds and squares decorated with playing-card shapes and curving panels and brackets. The carpenters also carved the Classical style's acanthus plant along the tops of windows. It all added up to a house which people came to think of as very English. (*Figure 10.1.*)

Richard Dale put a great deal of wood into Little Moreton Hall's extensions but the house was among the last to have wood used so lavishly. For thousands of years Britain had been covered by forests with many sturdy oaks and the supply seemed endless. But by the time work at the Hall began in 1559 many of the best trees had been cut down to make the charcoal that smelted iron for guns in British fighting ships. By the 1550s carpenters looked for methods of house building that took less wood.

During thousands of years there had been at least four main kinds of wooden house in Britain: Roof house; Cruck house; Jetty house; Box frame house. (*Figure 10.2.*)

1 Roof house

This sort had a roof but no walls. The roof rose from the ground. Tree branches leaned against each other either in a cone shape or in long rectangular huts partly held up by posts inside. After 1000 BC timbers were shaped by the adze and jointed and so planks were used for floors and walls under a thatched roof. This sort of house was still normal when the Romans took control of Britain in AD 43.

2 Cruck house

By AD 500 the Anglo Saxons had settled in Britain and they built wooden frames shaped like an A. These two-storey high A frames were two pieces of tree trunk crudely shaped and leaning against each other, bottom ends in the ground and top ends fixed together. This triangle of wood was called a **cruck**. Wooden walls and roof and the upper floor were supported by the cruck. People went on using this simple method until about 1600 when wood became scarce.

3 Jetty house

This method required smaller pieces of wood than the cruck method. The pieces were cut exactly to size for framing and they were spaced closely together. No upright piece was more than one storey high and the upper floor jutted out over the ground floor. This helped to keep rain off the walls below and to keep the upper floor beams straight. The overhang was a **jetty**. Houses were built with the jetty method from about the AD 1100s to 1500s. They could be made from less than

10.2 Types of wooden houses in Britain

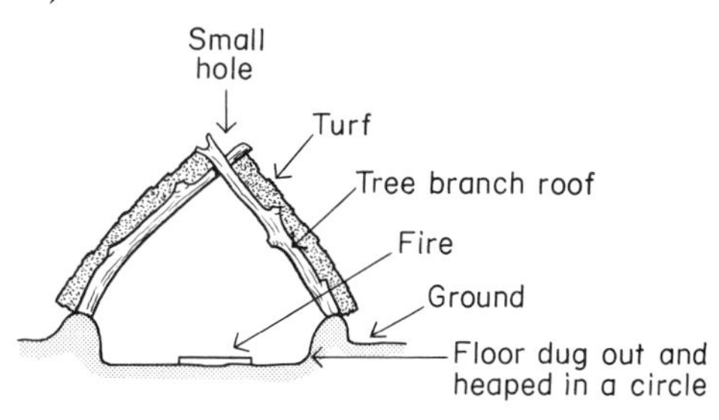

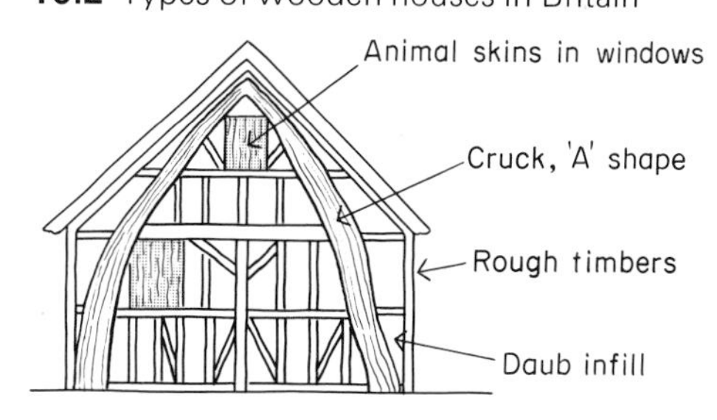

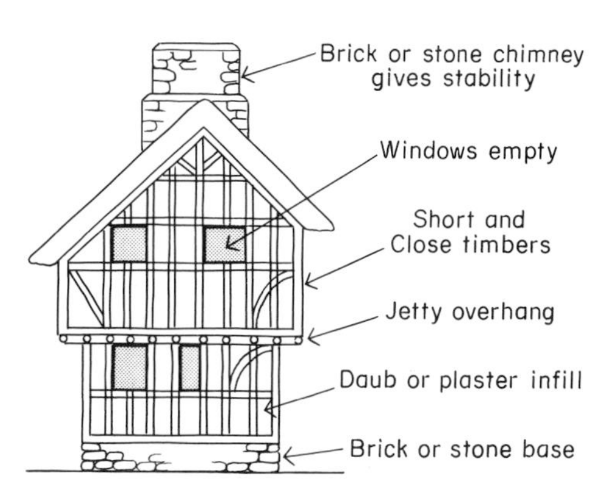

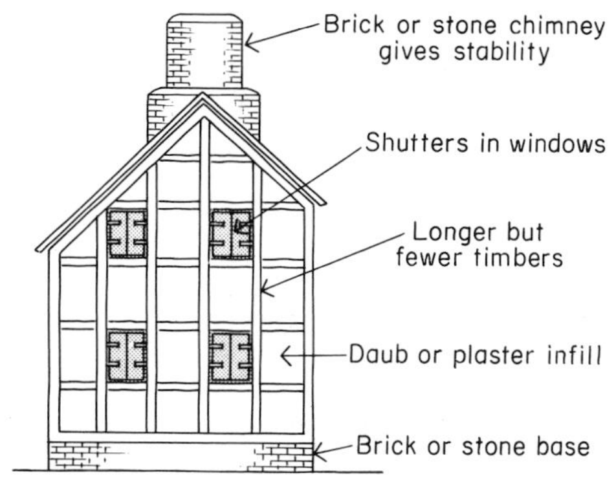

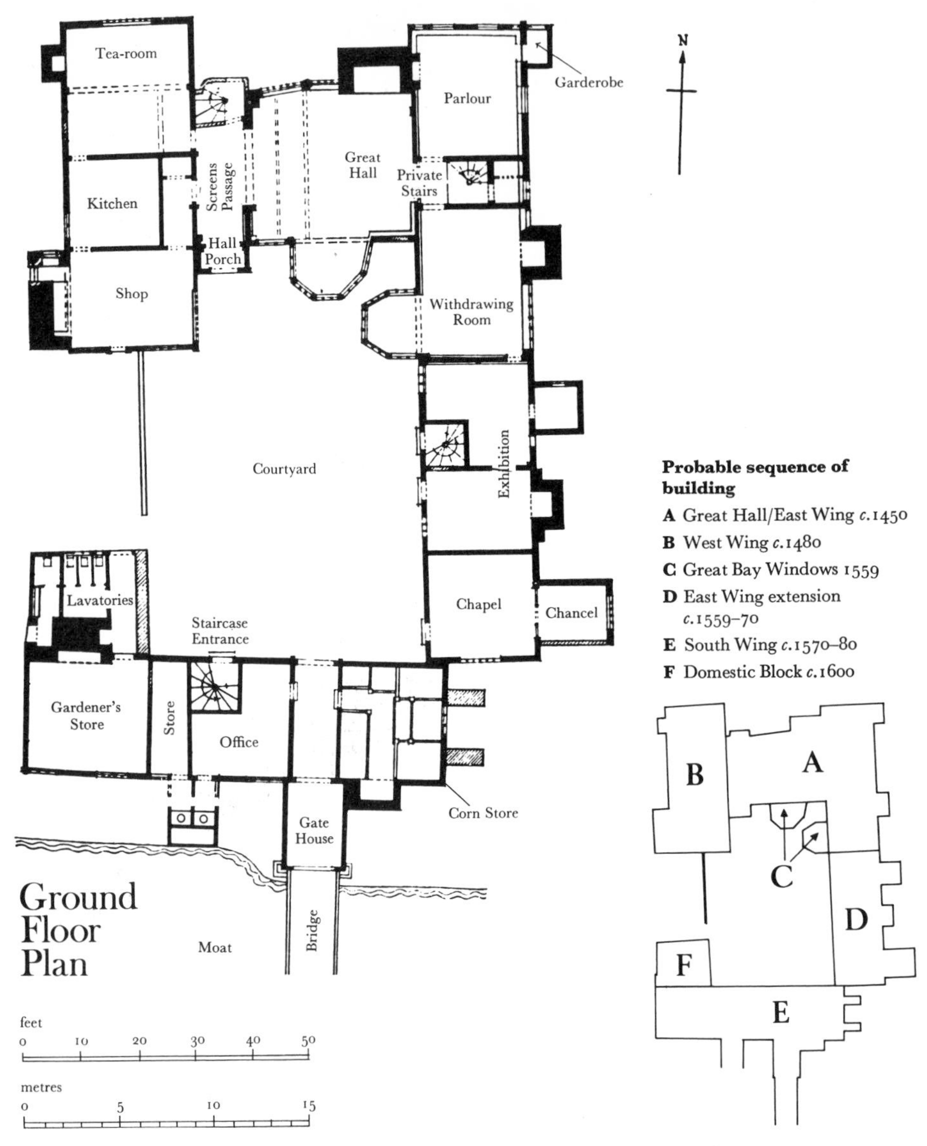

10.3 Plan of ground floor

perfect pieces of wood but they still used too much in a time of shortage. Little Moreton Hall was a type of jetty house.

4 Box frame house

This method was worked out by carpenters to save wood. So here the frame had longer pieces of timber and they were spaced further apart. Uprights went through two storeys and beams rather than short pieces were used for the horizontal parts of the frame. The box frame was a sort of wooden cage and it was first seen around 1500. Builders used it through more than 300 years, sometimes filling the voids with brick and putting a thin brick wall on the outside. By the 1980s a method something like this was being applied to small houses though the frames were insulated against heat loss and condensation.

William Moreton's son added another black-and-white room in 1580 to Little Moreton Hall and it ran the whole 23 m (75 ft) length on top of the 1559 gatehouse. (*Figure 10.3.*) This sort of **long gallery** was then being built in more and more country houses and it was to become a main feature of English stately homes.

Tudor houses like Little Moreton Hall looked back to the Gothic world which had gone before as well as forward to the Renaissance world which had only just arrived in England. A Gothic craftsman would have carved oak leaves and not the Italian acanthus leaves that Richard Dale chiselled along the Tudor windows of Little Moreton Hall.

PART 3 GOLDEN RECTANGLES

11 The Banqueting House, Whitehall

Place Whitehall, London
Time *Start* 1619
Finish 1622
Construction time 3 years
Purpose A hall for state occasions
Style Classical. English Palladian

The Banqueting House at Whitehall was England's first public building in the new Classical style from Italy where the style had begun in the 1420s. For a hundred years before 1600 people in Britain had been able to see fragments of Classical design in such details as doorways and chimney pieces in houses or on tombs in churches. But now in 1622 they could see a whole Classical building. (*Figure 11.1.*)

The Banqueting House was actually intended to become just a small part of a large new palace. The design drawings showed how the Palace of Whitehall would have had a front of nearly 400 m (1300 ft) along the River Thames and sides stretching back from the river nearly 300 m (1000 ft). The vast building was to be

11.1 The Whitehall front with Ionic Order below and Composite above. Rustication at ground floor and between windows

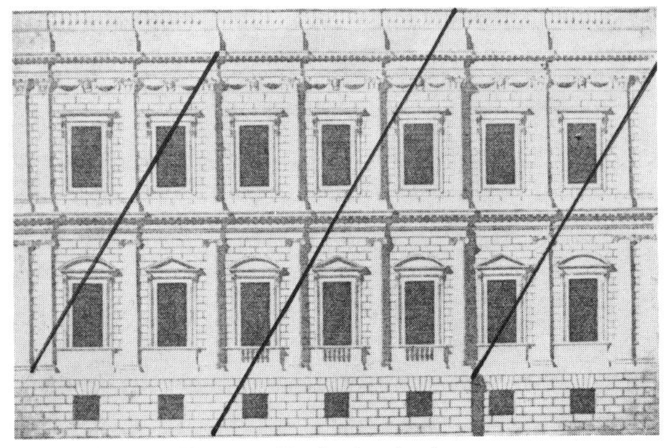

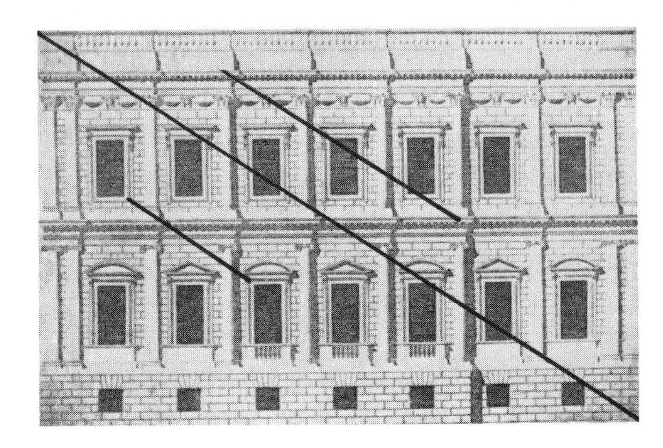

11.2 Golden Rectangles in the design

laid out as a grid of rooms and courtyards grouped around the Grand Court in the centre. The Banqueting House was to be tucked into the Grand Court's south-east corner. Yet out of all the hundreds of rooms planned for the Palace of Whitehall only one was built and that is the Banqueting House. And that one room began the 300 years of Classical design in Britain up to the Second World War in the 1940s.

By 1600 it had become normal for the sons of wealthy families to travel abroad as part of their education and Italy was the climax of their tour. They went to Rome and saw the 1500-year-old ruins and the vast new church of St Peter's with its dome designed by Michelangelo. They saw the city's new squares or 'piazzas' and the fountains. They travelled with their tutors to the splendid villas and gardens out in the country built by families as wealthy as their own but who lived among luxury and beauty undreamed of back home in England.

The young aristocrats went to Florence and saw the buildings by Filippo Brunelleschi who had been one of the first designers to use the Classical style again since Roman times. They saw the children's home he had designed in 1421 with its graceful row of columns carrying half circle arches which stood in front of the first building in the new Classical style. England did not get its first Classical building, the Whitehall Banqueting House until 200 years after this Foundling Hospital by Brunelleschi went up in Florence.

A clever 28-year-old artist from London first went to Italy in 1601. He was not wealthy but his visit was paid for by an aristocrat who was. The artist was Inigo Jones and he had gone to Italy to study painting but was soon studying the methods for the design and construction of buildings in the Classical style.

Among the new buildings Inigo Jones thought best were those by Andrea Palladio who had a style all his own as his country villas built in the 1550s around Vicenza showed. Palladio used Classical axis lines and Golden Rectangles and the Orders and all the other methods and features of the Classical style in a calm,

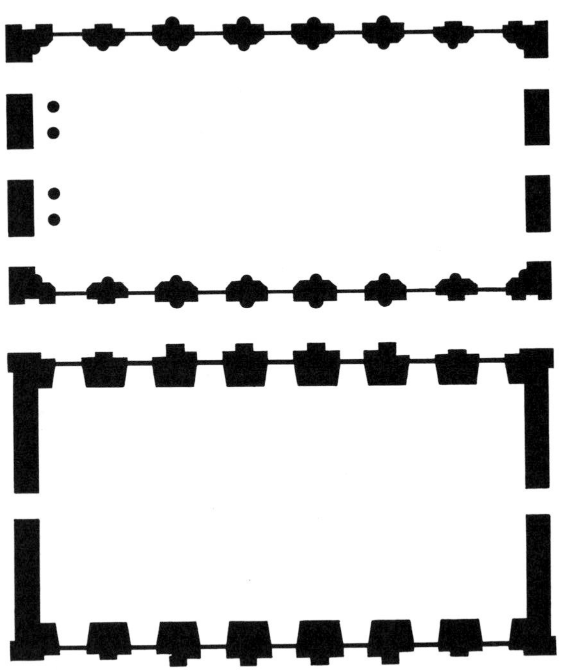

11.3 Plan

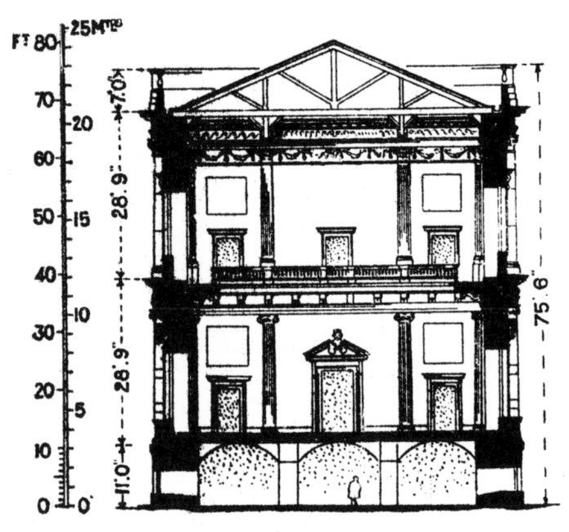

11.4 Section

restrained way that Inigo Jones liked. Inigo Jones brought that style to England. And it was that English Palladian style which became the main kind of Classical design in Britain.

Not many Englishmen could measure and draw ruins and buildings on site in Italy but they could study design books at home which was one of the main ways that Italian methods of design spread to other countries. Palladio himself had put his ideas into four design books and these were printed in Venice in 1570. Inigo Jones brought the books back to England. One of Palladio's designs showed a town palace and Jones used it when he built the Banqueting House at Whitehall between 1619 and 1622.

Another English designer, John Shute, had already been in Italy in the 1550s and he had his own design book printed in 1563. But in spite of this Jones was the first Englishman to see that Classical design was a system for bringing order to any building as a whole and to every part of that building. The basis of that order was symmetry and the basis of symmetry was geometry. For Inigo Jones the geometry was that of Golden Rectangles and their diagonals.

The Whitehall side of the Banqueting House was designed as a grid of Golden Eectangles. (*Figure 11.2.*) The overall shape of the building had its sides in the Golden ratio of 1 to 0·618, within that shape the position of the Orders and the windows and all the other features were fixed by an overlapping series of Golden Rectangles.

Inigo Jones designed the Hall as a double cube measuring 33·5 m (110 ft) long and 16·75 m (55 ft) high and wide. He put a narrow gallery half-way up and so designed the room's long walls inside and out as a two-storey building with two rows of Orders and two rows of windows. In 1635 the ceiling panels painted by Rubens were hoisted up and fixed. The glowing colours add richness to the white and gold of the room which is still today one of the grandest in England. (*Figures 11.3, 11.4 and 11.5.*)

Jones may have made this one-storey building look like two-storeys because he thought that one day the Banqueting House would have to fit in with the much larger but as yet un-built Palace of Whitehall. On the Whitehall side there was no entrance and so no portico. The entrance was at the north end and the one that is there today is so badly designed that it is no surprise to know that it was built by James Wyatt the 'Destroyer' who, when he died aged 67 in 1813, left a trail of professional vandalism behind him. In 1829 the designer John Soane restored all the outside stonework of the Banqueting House.

A short distance to the north at Covent Garden Inigo Jones built the small St Paul's church between 1631 and 1638 and he gave it a portico in a Golden Rectangle. But

11.5 Inside the white and gold hall looking at the Ionic doorway. Overhead are some ceiling panels painted by Rubens

at the other St Paul's, which was still London's old Norman cathedral on top of Ludgate Hill, he had the chance to build on the grand scale again. In 1633 he began a new portico at the cathedral's west end with eight Corinthian columns 12 m (40 ft) high and 1·2 m (4 ft) in diameter. He also cased the cathedral's crumbling sides with Portland stone blocks in the new Classical style.

Londoners had a shock when the scaffolding was removed in 1643 and they saw all this. But they had two even greater shocks just over 20 years later. In 1665 the plague swept through the city and brought the busy life of its streets and docks to a standstill. In 1666 the Great Fire of London followed the plague raging through the city's wooden buildings and then on to old St Paul's Cathedral which burned for two days and two nights.

Some 2·4 hectares (6 acres) of lead from the cathedral's roofs flowed down the streets in molten streams. Inigo Jones's new casing was split and his white portico cracked and blackened. Vaults and walls caved in. The cathedral stood up gaunt above the charred sites where 400 streets and 13,000 houses and hundreds of other buildings used to be. Thousands of people had escaped to camp out in the open country around the city.

Within a week the 33-year-old scientist, Dr Christopher Wren, was showing King Charles II a plan for a new City of London and for a new St Paul's Cathedral.

12 St Paul's Cathedral

Place Ludgate Hill, London
Time *Start* 1675
Finish 1710
Construction time 35 years
Purpose A church. London's cathedral and a national centre
Style Classical. English Baroque

12.1 Looking east through two arches of the central octagon

The building committee wanted an out-dated Gothic layout for the new St Paul's Cathedral and Christopher Wren wanted an up-to-date Classical plan. After several designs the cathedral finally got both. (*Figures 12.1, 12.2, 12.3 and 12.4.*)

Like the layout of most Gothic cathedrals St Paul's new building had two halls crossing at right-angles. Like Ely Cathedral's crossing which was finished in 1342, St Paul's crossing was also an octagonal space. But unlike Ely it had a dome, not a tower over it. Each of the two halls had vaults over the high nave and low aisles and flying buttresses outside to take the vault's thrust. At the west end stood two towers.

But the shapes Wren gave to this Gothic layout were in the Classical style called **Baroque** which designers were using in Italy by 1600. Baroque design was different from the earlier Classical style. Between 1420 and 1600 designers had liked to show all the parts of the building clearly as Gothic designers had done. But after 1600 Baroque designers ran one part into another making the parts less clear. They used curved outlines and surfaces to make a building look as if it were carved from one lump of rock rather than built of separate stones. Baroque designers liked to make buildings look as if meant for use by giants rather than by people.

The English Baroque style was plainer than Baroque design in other European countries. By the time St Paul's was finished in 1710 the Baroque style was no longer fashionable in England and designers went back to the even plainer Palladian style Inigo Jones had brought from Italy. But Baroque shapes and decorations were seen again in English design between 1900 and the start of the First World War in 1914 when it formed part of Edwardian design.

At St Paul's the Gothic layout was built up into a solid mass of masonry controlled by overlapping grids of Golden Rectangles. Inside these was a single giant Order with vaults on top and half-round arches between piers. Outside these were two Orders, one on top of the other like Inigo Jones' Banqueting House at Whitehall. (*Figure 12.5.*) These two Orders went right round the building and, between the two west towers, became the columns of a giant portico. (*Figure 12.6.*) Part of the upper storey outside was only a wall to hide the flying buttresses and to keep the block effect of the building simple. (*Figure 12.7.*) But high above the long and level skyline of the two-storey walls was the dome on its ring of columns. (*Figure 12.8.*)

Wren took his ideas for the whole of this central climax at St Paul's from the octagon at the crossing of the axis lines in Ely Cathedral. (*Figure 9.5.*). Wren knew Ely well and he studied the work there which was finished in 1342. Because Wren wanted a large space in the middle of St Paul's he placed an octagon there with huge piers at the eight corners. Wren joined his piers with half-circle arches and on top of them built a cylinder of stone nearly 30 m (100 ft) high. This cylinder was called the **drum** because of its shape. (*Figure 12.4.*).

The stone drum was 4·5 m (15 ft) thick. But it was a double drum and had an outer and an inner wall. Each wall was 1·2 m (4 ft) thick and the 2·2 m (7 ft) gap between the outer and inner wall was filled by a honeycomb of buttresses and arches giving extra strength with minimum weight. As the top of the double-drum's buttresses showed outside they were made to look like an Order. Seen from the streets below, this Order formed the ring of columns on which the famous outer dome appeared to sit. (*Figure 12.4.*).

The double-walled cylinder carried the whole dome system. From the top of the cylinder a cone of 460 mm (18 in.) thick brickwork was built nearly another 30 m (100 ft) high. And the top of this cone held up a stone lantern 27 m (88 ft) high which weighed 850 tonnes. And the top of the lantern carried a gilded copper ball and cross 108 m (355 ft) above floor level. (*Figure 12.4.*).

The half sphere of oak framework for the outer dome was built off the leaning surface of the brick cone. Under the cone was the inner dome also built of 460 mm (18 in.) brickwork. The whole dome system

12.2 Looking west from the central octagon under the dome

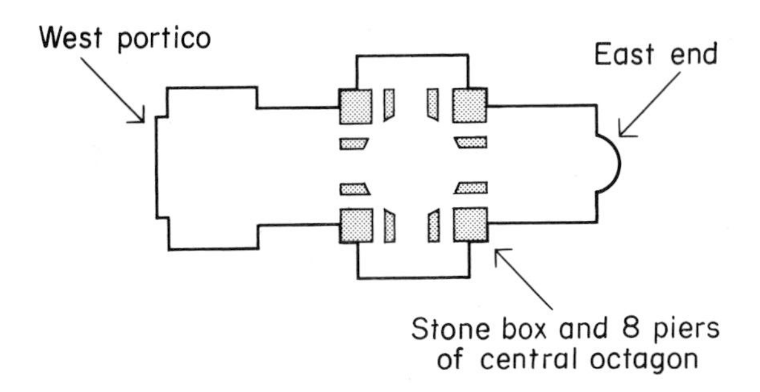

12.3 Plan

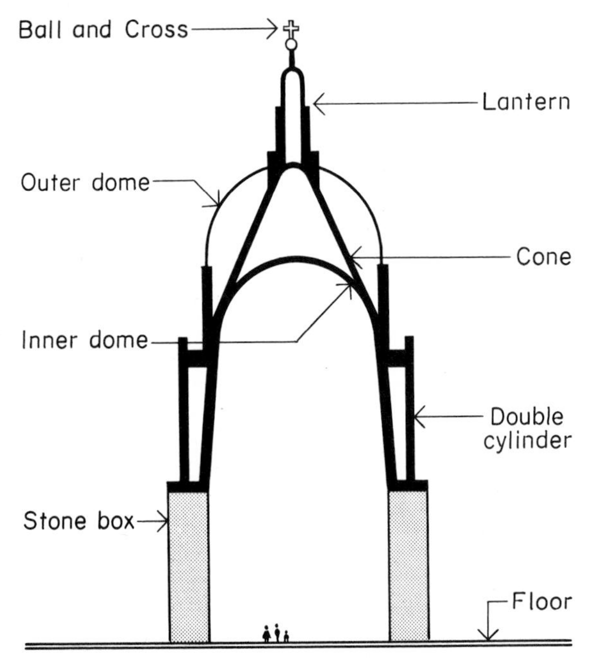

12.4 Dome construction

and the stone cylinder supporting it weighed over 23,000 tonnes, the weight carried by the eight piers. Extra support came from four corner bastions. Together with the piers they made a stone box with 8 m (25 ft) thick walls and sides 52 m (170 ft) long. Arches 28 m (90 ft) high opened the sides of the box to the spaces outside the central area.

The load on the foundations under the dome was 67,300 tonnes. No building in Britain had ever been so massive.

People on the site of the burnt-out old St Paul's soon found out how solid its Norman walls and piers had been because demolition of the ruins begun in 1668 went on for 19 years. Gangs of men used pickaxes. Some stood on top of the ruined walls, others worked from wooden platforms slung between huge wood scaffold poles. There were accidents. When a man fell the cathedral paid compensation.

The Norman walls were thickest near the ground. Pickaxes were too slow so Christopher Wren ordered a ram to be made. It was like a ship's mast with a steel spike on the end. It was slung from a frame and 30 men were needed to swing it to and fro. The ram broke and a second one was made 12 m (40 ft) long. It took two days to bring down only a small part of the old Norman tower, so a gunner was finally brought from the Tower of London. He put 8 kilos (18 lb) of gunpowder into a box and buried it in a hole under St Paul's tower. The fuse was lit. There was a loud rumble underground. The tower seemed to rise into the air and hang there for a moment before crashing to the ground. Four years later in 1672 nothing was left of the old Norman tower that had been a famous London landmark for over 500 years.

The site had to be controlled. Busy traffic passed up and down the streets all around so a wall was built with some of the old stones to keep people off the site. Huge piles of rubble had to be sorted. Stone that could not be re-used was carted away as fill on other city sites being rebuilt after the Great Fire.

Lead, copper, iron, marble and undamaged Portland stone were saved from the fire and, taking advantage of the low prices, the builders bought materials for the new cathedral. Storage space had to be found. Security guards with mastiff dogs patrolled the site at night. The guards were workmen and they received two-thirds of a day's pay for each night's duty. They were given warm cloaks to wear in winter and their dogs' meat was supplied free. The detailed records kept of this immense building project show that one of the dogs got mange and had to be taken to the vet.

Although there was such well-planned security there were still many thefts. Thieves stole wood, lead and a workman's tools. While building work stopped during the winter of 1704 the ends of lead sheets on roofs were cut off and stolen. Finally one of the masons was caught. Nails were a temptation for workmen too. They were easy to steal and good to sell. They were expensive as each nail was hand made by the blacksmith. One night vandals dragged some heavy ropes for scaffolding out and slashed them to pieces. Masons had to make the wall around the site much stronger and put spikes on top of gates to stop thieves entering.

But site security remained a problem. A public right of way ran across the site. People were supposed to keep to a walkway made of planks but nothing could keep the boys who lived nearby from using the site as a playground. They swarmed all over the place causing trouble and damage.

One man watching the building go up made the sort of remarks about the workmen that some people do today. He watched ten men moving a 153 kilo (3 cwt) block of stone on a trolley. He was sure the job could have been done by only two labourers, and all on public money!

Materials arriving at the site were checked in at the office-of-works by the main gate. The building committee did not want the public upset about public money being abused. Finance for the new cathedral came from many sources such as a tax on Londoners' coal. Boys stealing nails knew the old saying about 'Robbing Peter to pay Paul' which reminded people that Westminster's Abbey of St Peter's had some of its estates taken away to pay for repairs of the old St Paul's.

Every day horses and wagons came up the city streets with stone and oak from ships tied up along the River Thames. Quarry men down at Portland often quarried blocks weighing 10 tonnes and it took them years to find blocks of 20 tonnes. When loaded at Portland the ships sailed along the south coast but not all got to London. One sank and another was taken by French pirates. Those that got through sailed up the Thames as far as London Bridge. Their masts were too tall to go under the arches of the bridge so the Portland stones were off-loaded by cranes into barges and lighters.

Big blocks were hauled up the steep crowded streets by a capstan, turned by men or horses. Sometimes it took two weeks to get a big block through the city from river to site. Transport costs from the quarry to the site doubled the price of Portland stone.

In the masons' sheds on site the stone blocks were sawn up and shaped and the surfaces tooled. From scale working drawings or models supplied by Wren and his staff the joiners were busy making templates for the masons to shape their mouldings. Each mason owned a set of banker tools. For flat surfaces he had a **principal**. This was a broad sharp chisel which he struck with a mallet. He had other chisels and gouges for cutting mouldings and grooves. And a mitre tool for cutting angles in the stone. He had other tools for ruling, measuring and gauging as well as steel combs and rasps for making a perfectly smooth surface. It cost a great deal of money to get the hard Portland stones ready for their positions in the building. Softer stones, such as ragstone from Kent, could be cut quickly by the mason's axe.

Specialist craftsmen made ironwork or carved wood off-site at workshops down in the city. Their finished work arrived at the gate to be checked in. Two of the last horse-wagons were those which brought the ball and cross for the top of the lantern in 1708. Crowds watched as the ball went by in one wagon. It was of copper 1·8 m 6 ft) in diameter. And then the cross went by. This was of copper too, over 3·7 m × 3 m (12 ft × 10 ft). The crowds watched as they were hauled right to the very top of the new cathedral. Londoners at last could see their new landmark. (*Figure 12.8.*)

Christopher Wren not only designed St Paul's and all

12.5 The two-storey south wall with Orders. Part of the upper storey is only a screen wall hiding the flying buttresses. Above the skyline is the dome on its ring of columns

its details but he was also in charge of all the working drawings and models. And he was responsible too for the management of the project. There was no general contractor at St Paul's as there are on big building projects today. Contracts were given to different master craftsmen for certain parts of the building. Wren had to plan and co-ordinate all their work. In 1669 Wren became His Majesty's Surveyor General of the Royal Works as Inigo Jones had been before him. Wren was in charge of everything to do with royal and government building old and new as well as the royal quarries on the Dorset coast of Portland.

The Great Fire of London had added even more jobs to the long list being dealt with by the royal surveyor and his staff of technicians. Wren had to design more than 50 churches that had been burnt in the city fire though local craftsmen designed the detail of some. Those city churches became a training ground for craftsmen who would later work with Wren at St Paul's.

Contracts were drawn-up at St Paul's in three main ways:

1 Materials only

These were for the supply of materials only. The price was agreed before delivery.

12.6 The Orders outside become a two-storey portico with a pediment between the west end towers

2 Materials and craftsmen
These were for the supply not only of materials but also of the craftsmen to do a certain piece of work. Lead and ironwork was often done with this sort of contract. These were called **taskwork** contracts.
3 Craftsmen only
These were for the supply of craftsmen and unskilled men only. They used materials already stored on site. Some of these were **daywork** contracts.

Daywork contracts were the simplest kind. The contractor put an agreed number of men on a certain part of the work and he was paid a fixed sum for each day each man worked on site. The number of men might vary from day to day. Out of the money he was paid, the contractor took some for profit and paid his men out of the rest. Rates were fixed for each man's work as follows:

1 **The contractor** took the most.
2 **The craftsman** got 18% less than the contractor.
3 **The unskilled men** got 50% less than the contractor.

Contractors at St Paul's made a lot of money from daywork contracts. A roll-call was made three times a day to check the real number of men on site. This stopped contractors claiming for men who were not there. So at 6 am, 1 pm and 6 pm a bell rang and the men's names were checked. This job was done by the Clerk of the Cheque who also checked what was in the carts coming on and off the site. He got one-third extra pay for these two jobs.

All routine and daily work at St Paul's was done by these daywork contracts. Such contracts were for sawyers to cut up logs and for joiners to make patterns for the masons. Most carpentry was done by daywork, such as scaffolding, the making of centering for arches and vaults and the laying of wooden walkways across the site. Daywork contracts were also used for simple hoists which lifted stone and other materials. Erecting craftsmen's sheds and workshops was daywork, so was the construction of the cathedral's oak roofs and the making and fixing of wooden coverings to protect unfinished stone walls in winter when building work stopped. Wren liked to have daywork contracts for short-term work and also for detailed work where great care had to be taken.

The other sort of contract was for taskwork. There were three main kinds of taskwork contract:

1 Where the price was fixed by measurement after the work was done
This was the most usual kind of taskwork contract. The rates were fixed for a certain unit of work. For example, so much money for a certain unit of stonework. The final number of units in a wall or an arch was measured by the Assistant Surveyor or a trained clerk. If the work was below the agreed standard he could pay less for it.

2 Where the price was fixed before
This sort of contract was used for the demolition of old St Paul's as well as for fittings and ornaments in the new cathedral. Under this sort of contract the craftsmen often supplied their own materials.

3 Where the price was fixed by valuation after
Under these contracts the work was valued like a work of art. Before getting such a contract the craftsman had to be well known or give proof that he could do the work. He also had to trust in Wren's fairness as Surveyor. This kind of contract was only used for very special work such as fine carving.

The Great Fire of 1666 changed the whole building industry in London. Before the Fire there had been strict rules for any master craftsman wanting to set up his own business in the City of London. There was what in industry today would be called a 'closed-shop'. A close control was kept on craftsmen by the 'Guilds' which were something like today's trade unions and professional institutes. The Guilds were already old when the Great Hall was built at Westminster in the 1390s. A 'Freeman' was a person who had been given the right to do business and enjoyed other privileges

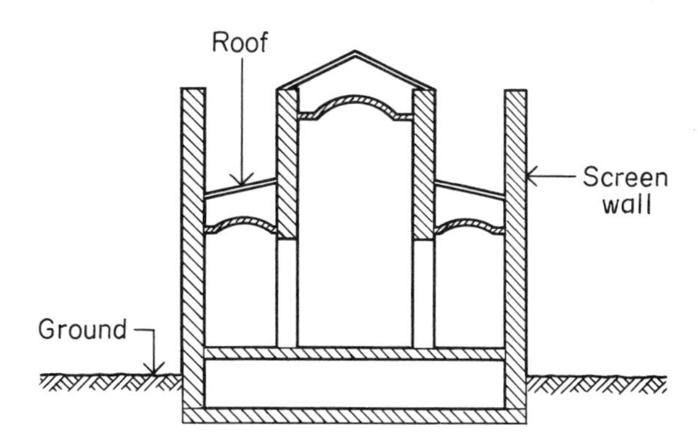

12.7 Section

that came with living and working in the City of London. Outsiders could get in but it was not easy.

London's building industry had been run by the powerful Guilds almost unchanged for hundreds of years. The City of London Guilds were also called **Livery Companies** and got their names after the dress or 'livery' their members were wearing before 1400. Many of the City Guilds still exist and the 12 which are still given privileges in the city today as 'Great' companies are: Mercers, Grocers, Drapers, Fishmongers, Goldsmiths, Skinners, Merchant Taylors, Haberdashers, Salters, Ironmongers, Vintners and Clothworkers.

Wren was very careful to keep on good terms with the City Guilds which had lost 44 of their famous Halls in the Great Fire and he encouraged members of the Guilds to join St Paul's Building Committee. Among the most regular attenders were John Cutler, grocer; John Frederick, barber surgeon; and the City Chamberlain, Tom Player.

Before the Great Fire there had been few big projects and most building firms were small. A master mason would employ about six skilled men called **journeymen** masons, and perhaps two apprentices and he worked with them himself. The master needed prompt payment for his work because he had little capital.

After the Great Fire the building boom changed the system. The old methods and rules of the City Guilds could not cope with the sudden need for thousands of homes and business places, together with public buildings, the Guilds' own Halls and churches as well as the vast project of the new St Paul's Cathedral. The boom drew in many master craftsmen from outside London who saw big money to be made from big contracts. All they needed was enough capital to buy large amounts of material and to pay the wages of many more journeymen than a master craftsman normally employed before the Great Fire. London had to relax its rules and there were plenty of men from outside London ready with money and skill.

12.8 A Second World War bomb-site view of drum, dome and lantern. At the top is the gilded copper ball and cross

Thomas Strong became one of London's new mason-contractors. He came from a Cotswold family well known as masons and he owned some big quarries at Taynton in Oxfordshire. After the Great Fire he went to London and in 1670 joined the Masons Guild under the emergency regulations and was made a Freeman of the City of London. Surveyor Wren gave him one of the two earliest contracts for stone walls at St Paul's. Thomas Strong began the work at the east end of the new cathedral just as the masons had begun at the east end of Durham Cathedral 600 years before. It was Thomas Strong who laid the foundation stone of St Paul's on 21 June 1675. As King Charles II was having trouble with the priests over Wren's design and with the government over finance for the new St Paul's, there were no fanfares or kettledrums. Everybody feared there would be riots if the King put on a big state occasion at the laying of the foundation stone for what was the grandest and most costly project of his reign. But Master-mason Strong was not without a sense of occasion himself and when he had laid the first stone he gave the trowel to Master-carpenter John Longland to lay the second stone. So this was how the new St Paul's was begun nine years after the Great Fire.

During the next 35 years hundreds of contracts, bills and accounts were stored in the site offices. Many of these records still exist and from them come the names of contracts and craftsmen who slowly, stone by stone, brick by brick, put up the immense building. Among them were:

Master masons Ephraim Beauchamp, Samuel Fulkes, Nathaniel Rawlins
Master bricklayers Richard Billinghurst and Thomas Warren
Master carpenters Israel Knowles, Richard Jennings, John James
Master joiners William Clear and Charles Hopson
Master coppersmith Andrew Niblett
Master plumbers Joseph and Matthew Roberts
Master plasterers Henry Doogood, John Grove, Chrysostem Wilkins
Founders Jane Brewer, Christopher Hodson
Carter and wharfinger John Slyford
Mercers Richard Turner, Nicholas Alexander
Sculptor Caius Gabriel Cibber
Damp-proofer Edisbury of Deptford

Christopher Wren knew them all and many became his friends. Some he outlived and some outlived him. Some had such faith in Wren's great design that when money ran short they lent their own savings. One of these investors was Mrs Shaw the proprietress of the Mitre Pub who for many years provided dinners for the master craftsmen.

When he died in 1723 aged 91 Wren was buried in St Paul's. A stone in the wall says that if anyone wants to see Christopher Wren's monument they must look around them. But Wren would also have wanted St Paul's Cathedral to be the monument of those who built it with him, even of old Father Smith, the organ builder. He and his assistant Bugbird accidentally set fire to the building on 27 February 1699 so that the new St Paul's nearly burnt to the ground like the old St Paul's before it. The crowd of craftsmen took two years to repair the damage.

13 Selfridge's

Place Oxford Street, London
Time *Start* 1907
Finish 1928
Construction time 21 years. But in two stages
Purpose A department store
Style Classical. Edwardian in the French style often called **Beaux Arts**

13.1 Anytime after 1928. Looking west along Oxford Street at London's shopping landmark. The 1907 first stage of the building is on the right

Around 1910 many designers were taking ideas and shapes from the dome and towers of St Paul's Cathedral. Edwardian designers liked its Baroque curves and carving. In 1908 Bramwell Thomas put a St Paul's-type dome on the new Belfast City Hall and put four Wren-like towers at the corners of the building.

The Classical style had been used non-stop in Britain during the 200 years that had passed since St Paul's was finished in 1710. But during that time the Classical style had gone through many changes, of which the Edwardian Baroque building in Belfast was just one example. The Edwardians liked another Classical style and it came from France. Design students at the School of Fine Arts – *Ecole des Beaux Arts* – learnt how to lay out their buildings with axis lines and grids of rectangles and how to make Classical buildings look very grand. One method was to use a 'giant' Order outside which went up the whole height of the building. The columns might be two, three or even more storeys high. This French style was called after the name of the Paris design school, *Beaux Arts*.

Selfridge's was a *Beaux Arts* design and the row of columns in its giant Order made the shop look like a palace. (*Figure 13.1.*) For reasons which today are hard to take seriously, people in 1907 liked a shop to look like a palace and not like a shop. But then they liked every sort of building from an abattoir to a zoo to look like a palace. But Selfridge's was more like a dream-palace than a real one because its stone columns were only decoration. The structural work was done by the building's hidden steel frame which Albert D. Miller had designed following the general idea for Selfridge's worked out by the American designer Daniel Burnham.

For more than 20 years Burnham had been designing large buildings in Chicago in the construction boom which followed the terrible fire there of 1871. Burnham knew how to use steel frames for fireproof multi-storey structures and he kept the outside of such buildings plain. Some Americans were also fond of dream-palaces but could not always afford them.

Steel did not burn but lost strength in the great heat when a building caught fire. So in America the frame's steel uprights called **stanchions** and the floor beams were covered in with fireproof blocks of clay tiles. The building's frame was seen outside as a grid of stanchions and beams filled in with large areas of glass. Some owners and designers of skyscrapers put dream-palaces at street level and at sky level but not in between. People did not take too much notice of a skyscraper's middle.

But Gordon Selfridge wanted his shop to be all dream-palace from top to bottom. And he knew what he wanted because not far away was a dream-palace built only three years before in 1904. This was the King Edward VII Galleries at the British Museum. John Burnet from Scotland designed these galleries in the French *Beaux Arts* style which he had learnt as a student at the Paris design school between 1874 and 1877. He put a giant Order outside. The galleries look like Selfridge's but without the shop windows.

Gordon Selfridge hired another American to design his department store in detail. This was the young Francis Swales who had also studied the *Beaux Arts* style in Paris. So Gordon Selfridge got a giant Order for his Oxford Street site. (*Figures 13.2 and 13.3.*) But the Selfridge's seen today was not finished until 1928. The final stages were dealt with by the same John Burnet who had designed the giant Order at the British Museum 24 years earlier. (*Figures 13.4 and 13.5.*)

By 1928 Selfridge's design was out of date, though it made sense for Burnet to finish the department store in the same style as Francis Swales had begun it in 1907. The design was out of date because after the First World War most owners could not afford lavish ornamental stonework outside their buildings and so plain styles were favoured as they were cheap to build. But there was a more basic reason why Selfridge's design was out of date. It had a built-in problem.

Any building in the Baroque or the *Beaux Arts* or any other Classical style had a heavy-looking ground floor of thick stonework. Windows were tiny. Inigo Jones's Banqueting House was like that, so was Christopher Wren's St Paul's Cathedral and so were John Burnet's 1904 galleries at the British Museum. But because a shop must have large windows at street level, the heavy ground floor at Selfridge's looked weak instead of strong. There were many other Classical buildings in Edwardian London where the necessity for a glass ground floor clashed with the Classical style's need for a stone ground floor. This was a design problem with no solution.

John Burnet had faced this problem long before he finished Selfridge's in 1928. Already by 1910 Burnet had designed the Kodak building in London's wide new street called Kingsway. He had gone to America to see

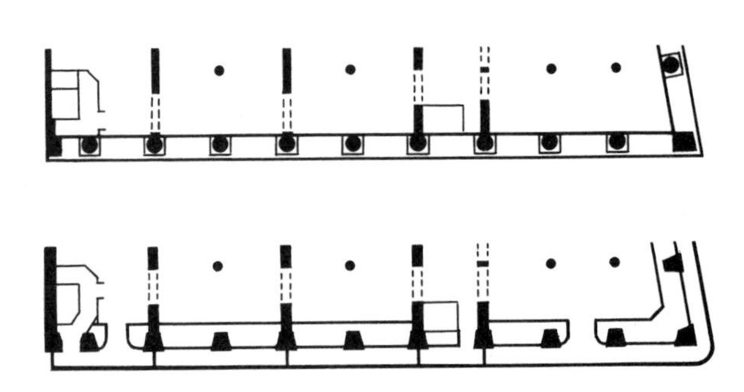

13.2 Plan of front at ground and first floors

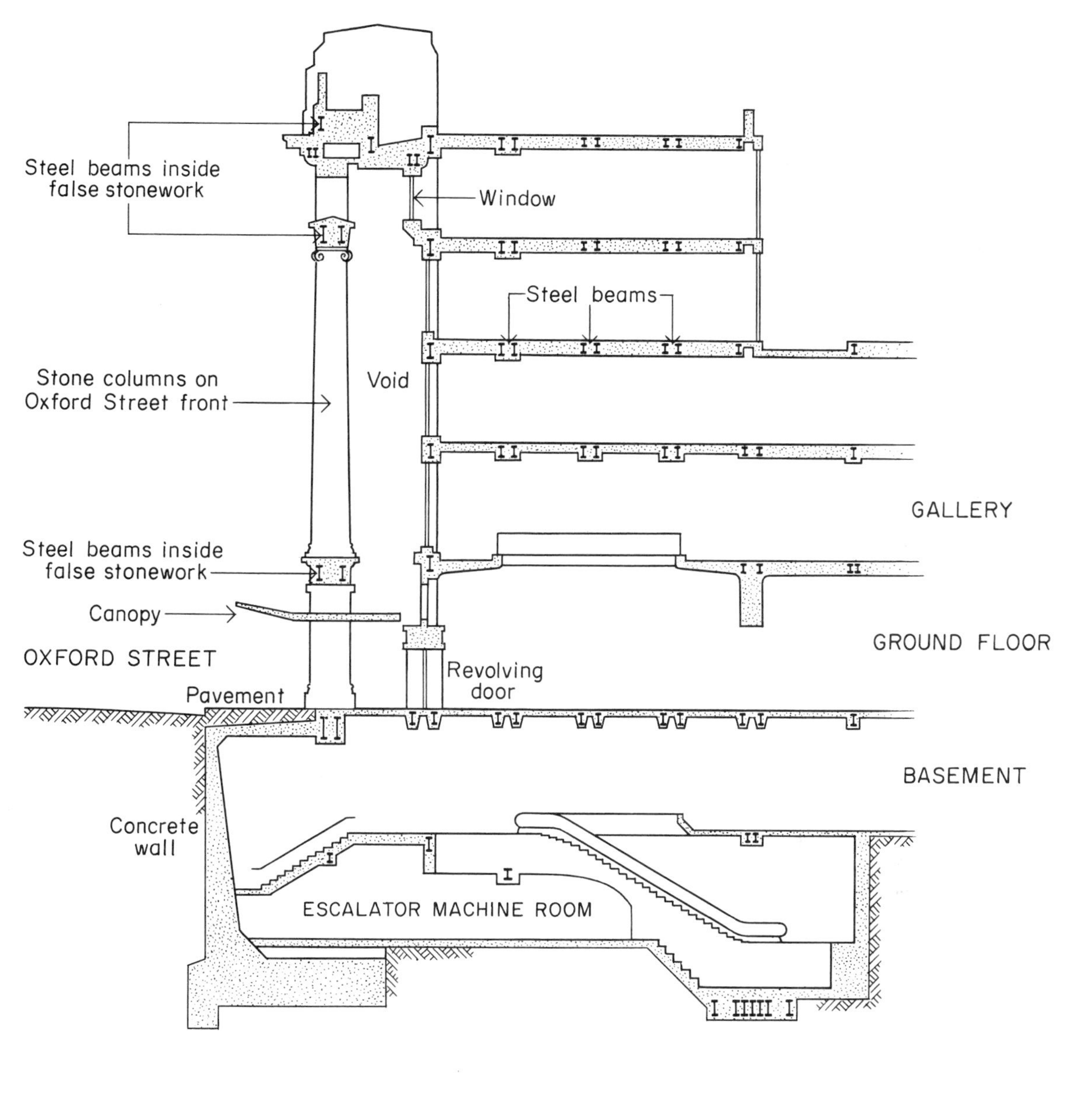
Steel beams inside
false stonework
Window
Steel beams
Stone columns on
Oxford Street front
Void
GALLERY
Steel beams inside
false stonework
Canopy
GROUND FLOOR
OXFORD STREET
Revolving
door
Pavement
BASEMENT
Concrete
wall
ESCALATOR MACHINE ROOM

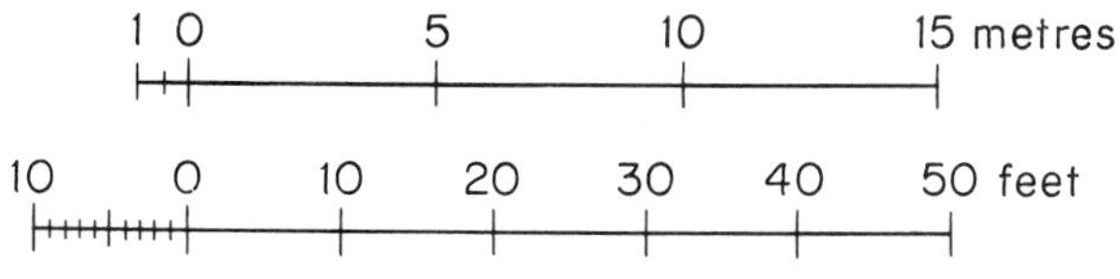
1 0
5
10
15 metres
10
0
10
20
30
40
50 feet

13.3 Section

13.4 About 1910. The first stage of the new building is on the right

13.5 About 1925. Only the middle of the site still has to be cleared of old houses

13.6 Kodak House, Kingsway

how up-to-date office buildings were being designed there. Kodak House was the result of what he saw across the Atlantic. It had a steel frame and large ground floor windows and looked like a real building, not a dream-palace. Kodak House only took a year to build and was one of the first buildings in Britain that could be called **Modern**. This was because its steel frame did not hide behind the fancy-dress of the Classical style or the Gothic Revival style or any other style from the past. Kodak House was, and still is, one of London's best buildings. It did not and still does not get the praise it deserves.

The 1907 Selfridge's was at the end of the Classical style and the 1910 Kodak House was at the beginning of the Modern style. (*Figures 13.6 and 13.7.*) Yet even the steel frame at Kodak House had a precedent. And that went back more than a hundred years to the first metal-framed building in the world which was a textile mill built at Shrewsbury in 1796.

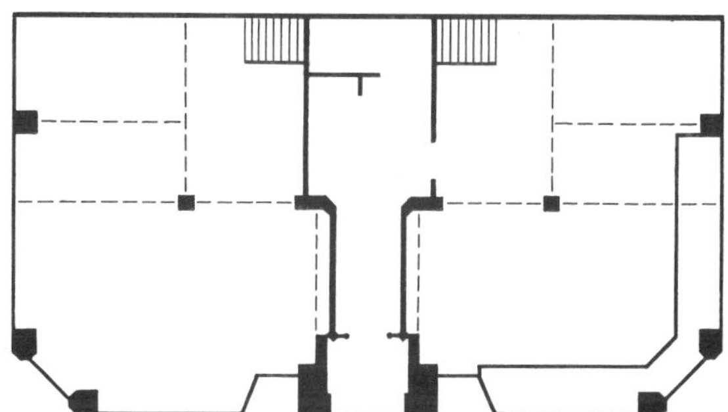

13.7 Plan
Designer John Burnet

PART 4 IRON RULES

14 Bage's Mill

Place Shrewsbury, Shropshire
Time *Start* 1796
Finish 1797
Construction time Just over 11 months
Purpose A textile factory, known as a **mill**
Style Georgian Industrial

14.1 Rows of columns inside the mill

By the 1700s big warehouses existed in most large towns where there was access to the sea or a river or later to a canal. The thick brick or stone walls of the warehouses supported 300 mm (12 in.) square wooden floor beams which carried the heavy loads of goods stored inside. The buildings often had five or six storeys each about 3·5 m (10 ft) high and about 8·5 m (28 ft) wide. Many had wooden posts inside to prop up the heavily loaded beams.

Flour mills were buildings similar to warehouses standing near fast-flowing water which turned the flour-milling machinery inside. When machinery for spinning and weaving was invented this machinery too went into mill-like structures near fast-flowing water. In 1718 John Lumbe built a mill near Derby with a 5·5 m (18 ft) diameter water-wheel to drive the 26,000 wheels of the mill's silk-making machines. After 1765 steam engines took over from water power so mills then needed to be near coal mines as there was no quick or cheap way to move coal from the mines to the mills.

Fire hazard was the main problem faced by the owners of all these warehouses and mills. The goods inside caught fire when workers knocked over candles or oil lamps. In textile mills the wooden floors were always soaked in oil which dripped from the machines. The mill owners did not like fires. They lost valuable raw materials and finished goods and they lost profits while new machinery was being put into new buildings. Of less concern to the mill owners in many cases was the loss of human life. They knew there would always be a supply of men, women and children to tend the machines in place of those who were burnt or killed.

Because of so many fires, insurance premiums were very high. Between 1791 and 1792 fire gutted five mills near Derby alone and the worst of all was also in 1791 when the Albion flour mill in London burnt to the ground. The Albion mill was a new building full of the latest machinery. Nothing was left. There was panic. The insurance companies either refused to insure mill owners at all or made the premiums so high as to threaten their business.

It was clear to everybody that a method had to be found of building fireproof mills – and found quickly. So it was loss of money rather than loss of life which led to Britain's first buildings with fireproof floors and to the first buildings with metal frames. The floors were brick. The frames were iron. But the ideas were French.

Like British mills French theatres were always burning down and for the same reason. Candles and oil lamps set fire to the costumes and scenery and they in turn set fire to the wooden buildings. Galleries and roofs fell in so that even the famous French comedies ended as tragedies. In 1790 the Royal Palace Theatre in Paris burnt down. The designer Victor Louis rebuilt it and only used materials which would not burn. To make the new theatre's roof fireproof he put two ideas together.

First he used wrought iron trusses like those which spanned 16 m (52 ft) over a staircase at the Louvre Palace in Paris. The designer of these trusses, Jacques Soufflot, had built them in 1780 and it was one of the world's first iron roofs. It had glass on top to light the stairs. So iron trusses went over the Royal Palace Theatre.

Louis's second idea was to make the roof's sloping surfaces with hollow clay blocks set in a sort of concrete. He got this idea from the fireproof floors built by another Frenchman Eustache St Far in 1785. A different kind of fireproof floor had already been used in 1741 by the designer Contant d'Ivry for a house at Vernon in the south of France. He built low brick barrel vaults between floor beams, the idea coming from buildings he had seen in Spain.

By 1790 these up-to-date French ideas were known in London. In 1791, the year of the Albion mill fire, a professional report said that the use of hollow clay tiles or **pots** was a good way to stop fires. And in 1792 John Soane designed some simple hollow pots for building a fireproof dome over a new wing at the Bank of England. At the top he put a round opening covered by an iron frame filled in with glass. He used no wood at all in this dome.

In that same year William Strutt took up these methods for the fireproof multi-storey mill he built at Derby. And it was from Strutt that Charles Bage got some of the ideas he used in his 1796 mill at Shrewsbury. (*Figures 14.1 to 14.5.*)

Charles Bage was born in Derby but in 1780, at the age of 28, he went to live in Shrewsbury and became a wine merchant. He was interested in engineering, keeping up with the latest ideas. He even had ideas of his own for using iron as a basic building material. For at least a hundred years Shrewsbury had been the world's main centre for new methods of making and using iron.

Iron had been used in buildings for thousands of years but only in small pieces such as nails or cramps for joints in stonework. Michelangelo put great iron chains round the dome of St Peter's in Rome which was finished in 1590. Wren put chains round St Paul's dome which was finished in 1708. (*Figure 12.4.*) At the new east wing of the Louvre Palace in Paris the designer Claude Perrault held up much of the stonework by slotting the stones round cages of wrought iron. By this method a number of small stones could be made to span from the top of one column to the next instead of one big stone. Jacques Soufflot also used iron cages at the domed church now called the 'Pantheon' in Paris. Its layout was like St Paul's in London but Soufflot did not

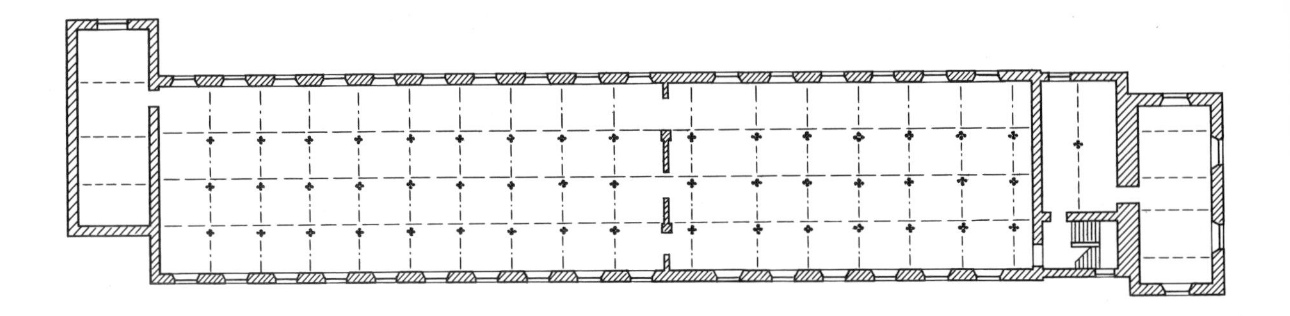

14.2 Plan of typical floor

want heavy piers in the central area like St Paul's which blocked people's view. So instead of piers Soufflot used thin stone columns to hold up the dome. To make the columns strong enough he put wrought iron cages between the joints.

The size, shape and position of the iron cages were worked out by Emiland Gauthey using methods from mechanics and mathematics. Gauthey also tested the strength of the stones being used for the thin columns supporting the Pantheon's dome and the machine he made for testing the samples was a new idea. These were the first scientific tests ever done on building materials.

Iron had never been easy to make. Iron ore had to be burnt in furnaces at 1530°C until the iron melted and ran out. Great heat needed great amounts of fuel. People had tried to find ways of getting more heat for less fuel.

In 1709 Abraham Darby invented coke at Shrewsbury. He burnt the gas out of coal by using a coke oven. Coke gave great heat and not only made better iron but also more iron than coal. By 1750 most foundries were using coke. Britain's growing industries with their machines needed more and more iron. By 1820 Britain's foundries were pouring out nearly 600 times as much iron as in the 1740s.

Coal was needed for making the coke that melted the iron. Coal was also needed for the steam engines which worked the great bellows for blasts of air to heat the coke to 1530°C. And coal was needed for the steam engines which lifted the heavy mechanical hammers that **forged** wrought iron. Coal came by horse-drawn barge to Shrewsbury's foundries when its canal opened in 1798. And by the 1830s coal was moving about Britain along its 3200 km (2000 miles) of iron railway lines. Many new bridges had to be built to take the new railways.

Shrewsbury's foundries made iron in three main stages:

1 **Pig iron** This was the molten iron which flowed out when ore was burnt in the furnaces. The white hot iron ran out into channels in the foundry floor. It cooled into blocks called **pigs**. Pig iron was the raw material for cast iron and for wrought iron.

2 **Cast iron** The pig iron was re-melted and run out or 'cast' into moulds so that the cooled iron or 'casting' took the mould's shape. Wooden patterns pressed the wanted shape into special sand as a mould for the castings. The wooden pattern and the sand could be used again and again so that any number of castings could be made all the same size though only in shapes which sand could take easily from the wooden patterns.

Cast iron was strong when squashed by heavy weights and so was good **in compression**. But it was weak when bent or pulled by heavy weights. Cast iron was not good **in tension**: it snapped easily and was brittle.

3 **Wrought iron** The pig iron was re-melted but also mixed with other materials which were stirred together in their molten state in a **puddling** furnace. The pig iron and other materials fused together as a white-hot dough which was removed from the furnace and hammered. By the 1700s this **forging** process was done by massive iron hammers lifted by water-wheels and later by steam power.

The forging hammer squeezed out unwanted material and what was left was 'wrought' iron. Wrought iron weighed only half as much as cast iron and could be pulled or bent without snapping and as it was good in tension it could be used in bridges and buildings. At first only small amounts of wrought iron were made because it cost more than cast iron.

Many new methods of making iron were being tried out by foundries in the Shrewsbury district when Charles Bage went to live there in 1780. In 1779 the famous Ironbridge had been built across the steep banks of the River Severn 20 km (12 miles) from Shrewsbury and Abraham Darby made the cast iron parts for it at his Coalbridge foundry and it was the world's first bridge made wholly of iron. In 1789 cast iron columns were used to hold up the gallery and roof of a church at Wellington and in 1792 St Chad's Church in Shrewsbury itself got two tiers of cast iron columns 11·5 m (38 ft) high. And then William Hazeldine set up a new foundry in Shrewsbury. Four years later in 1796 he was to cast the iron columns and beams for Bage's mill by the new canal.

There was a good deal of other construction work with iron going on around Shrewsbury. In 1788 Thomas Telford had become County Surveyor and he was the new sort of designer who understood how iron was made and how it could be used and he knew how to work out new building methods by mechanics and mathematics. He designed the cast iron bridge to carry the new Shrewsbury Canal over the River Tern at Longdon. The canal barges went over the river in a 1·4 m (4·5 ft) deep cast iron trough filled with water and the Longdon canal bridge acted as a deep beam supported by 2·75 m (9 ft) high cast iron columns.

Telford worked out the Longdon design with William Reynolds of the Ketley foundry in the Coalbrookdale area. Charles Bage watched the casting and construction of Telford's bridge which was finished in 1796, six months before he began his mill, and he also had notes about the results of tests carried out in 1795 on iron columns for the canal bridge.

Bage worked out that an iron column 3 m (10 ft) long would carry 28·5 tonnes before breaking if it was cast with a 400 mm (4 in.) X Section 25 mm (1 in.) thick. Iron columns in the lower storeys of his mill would each have to carry about 40 tonnes. His method showed that iron could carry 0·6 tonnes per sq cm. In 1862 John Rankine used another method and got the metal's strength as 0·4 tonnes. Today's more advanced methods put iron's carrying strength somewhere in between and it varies with different castings. So Charles Bage came very close to today's methods.

For over 300 years people had been trying to discover scientific facts about the best shapes for structures, the effect of weight on beams and columns and the strength of materials. The famous Italian inventor and artist Leonardo da Vinci (1452–1519) made hundreds of notes and sketches regarding the problems of designing beams and columns, arches and trusses. He did experiments and used mathematics. He also had ideas about using iron for bridges, but it was 300 years before the world's first iron bridge was built across the Severn in 1779 at Ironbridge.

The Italian Galileo (1564–1642) was helped by Leonado da Vinci's notebooks when he studied tension in structural members. Galileo also looked at what effect the bending of a beam had on the material inside the beam. His ideas and methods became the starting point for all later work on structural mechanics.

The Englishman Robert Hooke did some experiments in 1678. With wire and springs he found facts about the way materials changed shape when loaded. And in 1705 Jacob Bernoulli had ideas about the way beams sagged when loaded. This sag is called **deflection**. Bernoulli's work is still useful today. In 1757 the Swiss mathematics expert Leonard Euler made his ideas known about the strength of columns. He showed designers how to work out the way a column would bend sideways under load. This is **buckling**. Euler's ideas were the starting-point for all later work on column buckling. And in 1773 C A Coulomb at last got at the true facts about tension and compression in beams.

Two schools for structural engineering opened in France. The School of Bridges and Roads in 1747 and the Polytechnic School in 1795, the year before Charles Bage designed the iron beams and columns for the Shrewsbury mill. The French teachers and students spent a great deal of time over mathematics and structural mechanics. They liked to work things out in their heads and on paper.

Britain had nothing like these two schools and war between Britain and France meant that ideas from France were slow to cross the English Channel. In some ways British ideas were 50 years behind the French. Yet British designers found their own way to many of the ideas about structural mechanics. And in Britain people were beginning to use iron for buildings. They liked to design by testing sample columns and beams. They tested them by loading the sample castings until the iron broke under the weight. This was the sort of test done at the Ketley foundry which was so useful to Bage.

Charles Bage's new mill began on site in October 1796. It was a simple brick building 54 × 12 m (177 × 40 ft) and 16·5 m (54 ft) high in five storeys. The bottom and the top two storeys were 3·3 m (11 ft) high while the second and third were 3·25 m (10·75 ft) and 2·9 m (9·5 ft) respectively. The brick walls were 584 mm (23 in.) thick at the two lowest storeys and reduced to 330 mm/13 in. at the top storey. The walls were like those of any mill. It was the grid of cast iron floor beams and columns inside that made Bage's mill different. This was the world's first iron-framed structure. It had three main parts: columns, beams, fireproof floors.

1 **Columns** Three rows ran down the whole length of the building on each floor. The rows were 2·9 m (9·5 ft) apart and supported the iron floor beams. The columns had an X section and those at the lowest storeys were 150 mm (6 in.) across with an area of 181·5 sq cm (29 sq in.).

Their shape swelled out in the middle. Bage knew from the design rules he had worked out for cast iron columns that they would be most liable to bend in the middle, so he thickened the cast iron there. Bage was so accurate that these columns would be almost the same if designed by today's methods.

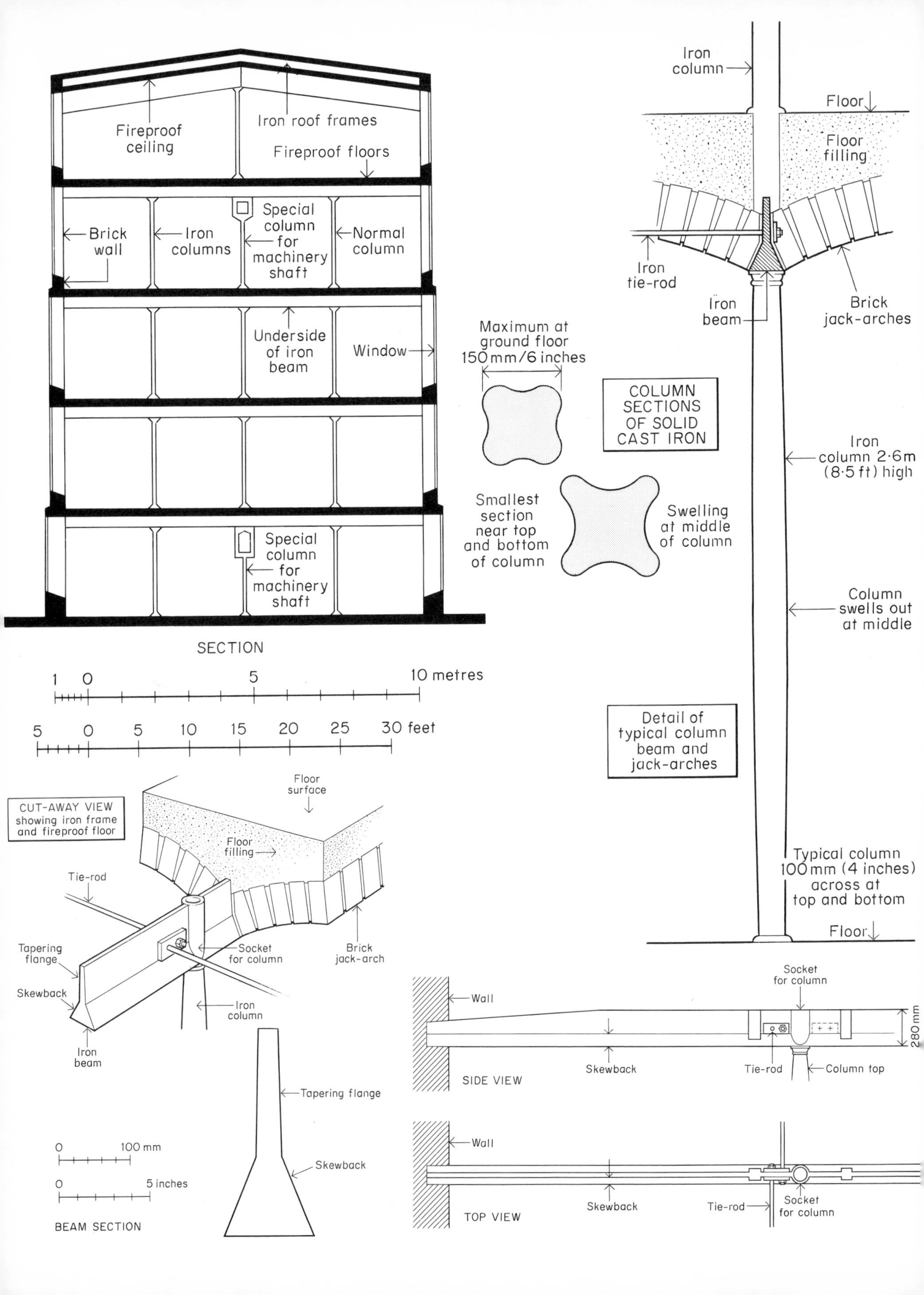

Fireproof ceiling
Iron roof frames
Fireproof floors
Brick wall
Iron columns
Special column for machinery shaft
Normal column
Underside of iron beam
Window
Special column for machinery shaft
SECTION
1 0 5 10 metres
5 0 5 10 15 20 25 30 feet
Iron column
Floor
Floor filling
Iron tie-rod
Iron beam
Brick jack-arches
Maximum at ground floor 150 mm/6 inches
COLUMN SECTIONS OF SOLID CAST IRON
Iron column 2·6 m (8·5 ft) high
Smallest section near top and bottom of column
Swelling at middle of column
Column swells out at middle
Detail of typical column beam and jack-arches
Typical column 100 mm (4 inches) across at top and bottom
Floor
CUT-AWAY VIEW showing iron frame and fireproof floor
Floor surface
Floor filling
Tie-rod
Tapering flange
Socket for column
Brick jack-arch
Skewback
Iron column
Iron beam
Tapering flange
Skewback
0 100 mm
0 5 inches
BEAM SECTION
Socket for column
Wall
280 mm
Skewback
Tie-rod
Column top
SIDE VIEW
Wall
Skewback
Tie-rod
Socket for column
TOP VIEW

2 Beams These went 11 m (36 ft) from wall to wall and rested on the columns at every 2·9 m (9·5 ft). Here they were cast as an upright tube shape or **socket** to fix the columns. The beams were cast in two 5·5 m (18 ft) halves for ease of casting and transport from William Hazeldine's foundry. They were bolted end-to-end through a flange.

Where they rested on the outside walls the beams were 177 mm (7 in.) deep and 280 mm (11 in.) halfway between column tops and 254 mm (10 in.) over the column tops. Bage knew that the beams would have to be deeper where they were more liable to bend under the heavy loads of the floor and the machinery. He knew the beams would have to be strong enough to resist these **bending stresses** and he knew in what parts of the beam those stresses would have their greatest and their least effect.

As design methods later became even more accurate than Bage's, mathematics and facts about the strength of the materials could be used to find the exact location of bending stresses inside a beam. This was vital for the design of structure in reinforced concrete where steel rods were placed in those parts of a beam subject to bending stresses.

Bage's cast iron beams were not a material ready made by nature. They did not exist until they were cast at the foundry. They could not be cast until they were designed and they could not be designed until Bage knew where the bending stresses would be so that the iron could be cast to the right thickness.

Nobody had ever done for a building what Bage did. He used a man-made material which had no natural shape of its own. And in order to give it shape he had to use scientific methods of design; so his complete iron framed mill at Shrewsbury is the first modern structure.

3 Fireproof floors Bage's floors had **jack arches** between the beams like the method first used at Vernon in France 50 years before. The bottom edges of each jack-arch came at an acute angle against the side of the beam. So the bottom part of the beam was thickened out with a sloping surface called a **skewback** for the jack-arch to rest on. (*Figures 14.3, 14.4 and 14.5.*)

Bage made his skewbacks as part of the castings. The bottom surface of Bage's beams was 127 mm (5 in.) wide narrowing inwards to 39 mm ($1\frac{1}{2}$ in.) and narrowing again to 32 mm ($1\frac{1}{4}$ in.) at the top.

To keep the beams and columns upright while the heavy jack-arches were being built off the iron skewbacks; tie-rods of iron were fixed from beam to beam. The top of the jack-arches were filled in and levelled off with a sort of concrete.

14.3 Cross-section and details

14.4 Details of columns, beams and fireproof floors

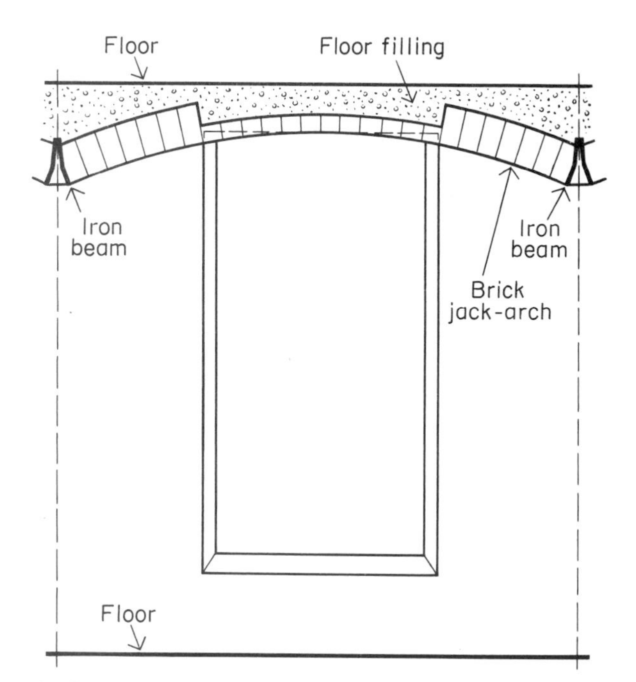

14.5 One bay of fireproof floor

Bage went on to design and build at least two other mills. His rules of design for iron structures led him to a new kind of beam in the mill he built at Meadow Lane in Leeds in 1802. By that time Bage had replaced the skew back with a simple cross piece or 'flange' at the bottom of the beam. He knew that this cross piece was enough to help the cast iron resist the stretching action or 'tension' due to the bending stresses. So Bage's beam with one flange was the first move towards the double-flange I section used today.

The people of Shrewsbury seemed to have been pleased with the new mill which they knew by the names of the business partners, Marshall, Benyons and Bage's Mill. They had watched it go up in just over eleven months and saw the 20 hp steam engine arrive in parts from the makers Boulton and Watt. It is not known now if the townspeople noticed that while designer Bage had kept to the Iron Rules he had not kept to any rules of the Classical style that reigned over design just as King George III reigned over the people in 1797. But the style, sometimes called **Georgian**, not only meant charming houses with white-painted doorways and marble fireplaces, but also industrial buildings such as windmills and watermills or textile mills like Marshall, Benyons and Bage's.

15 The Palm House, Kew

Place The Royal Botanic Gardens, Kew, Surrey
Time *Start* 1844
Finish 1848
Construction time 4 years
Purpose A glass-house for tropical plants
Style Victorian iron-and-glass

15.1 The see-through building across the lake

By 1800 many types of tropical plants had been brought to grow in Britain. Shrubs and flowering trees, palms and orchids needed a great deal of light and warm, damp air. The first glass-houses built for them were wooden structures. Thin wooden frames and small panes of glass let the light in but, although the hot and humid air was good for the plants, it was bad for the wooden frames which soon rotted. In 1815 G S Mackenzie had the idea of building glass-houses with iron frames and in 1818 the garden expert and journalist John Loudon did some sketches of curved glass-houses.

In 1822 Loudon wrote about a 14 m (45 ft) high glass-house and another that was 30·5 m (100 ft) long. Also in Yorkshire there was a heated glass dome 18 m (60 ft) high by 1829 and in Sussex people could see one at Hove by 1832. This was the 'Anthaeum' designed in the iron-and-glass style by Amon Henry Wilds.

The Anthaeum was the world's largest iron-and-glass dome with a diameter of 49 m (160 ft) and a height of 20 m (64 ft). When it was finished the Anthaeum was to have had a lake and a hill made with rocks as well as tropical plants. There were to be birds in the trees and high up in the dome a circular viewing gallery 8·3 m (27 ft) across with an observatory in the middle. All these delights were to be enclosed in the glass dome with curving ribs of cast iron that were fixed to iron plates on top of the foundations 3 m (10 ft) below ground.

The iron-and-glass dome went up but it was neither designed nor built properly and an hour after the last workman left the site the whole structure fell down. The head gardener was under the dome tending some of the plants when he heard a loud crack. He ran outside and jumped over the boundary wall just as the vast bubble burst, to lie in a tangle of shattered iron and glass for 20 years before being cleared away. The shock of the disaster sent the botanist Henry Phillips blind for he had been one of the main organisers. The Anthaeum was never rebuilt.

The Duke of Devonshire had a seaside house in Brighton and along with other local people he had given some tropical plants to the ill-fated Anthaeum. But at his stately home, Chatsworth House in Derbyshire, he had better luck when his head gardener Joseph Paxton got together with the designer Decimus Burton in 1836 to design the Duke's new Great Conservatory. Paxton had many ideas for iron-framed glass-houses. Both men knew what garden-journalist Mackenzie had said about iron frames and they knew that T A Knight had used iron for a glass-house at Downton Castle in Shropshire where so many iron structures had already been built.

So the 84 m (275 ft) long Great Conservatory rose from the ground at Chatsworth with two curving surfaces of glass down the whole of each side carried by curving cast-iron frames. Despite the risk of rot Paxton fixed the glass with thin wooden sash bars made in a cutting machine he designed for the job. It would have taken joiners too long to make the thousands of curving sash bars by hand, and even by the machine method the Great Conservatory took four years to build.

Not many people could travel abroad in the 1840s and it thrilled them to see tropical plants growing in glass-houses. But people in Ireland were the first to see an iron glass-house specially built to hold palm trees. They could walk around the Botanic Gardens of Belfast in 1839 and see the iron structure designed by Charles Lanyon and cast at Richard Turner's foundry in Dublin in which city the glass was also made. Turner put up other structures in Ireland like the Belfast palm house. There was one at Portlaoise, another at Glasnevin, and a third at Killikee.

The design of the Palm House or Palm Stove for the

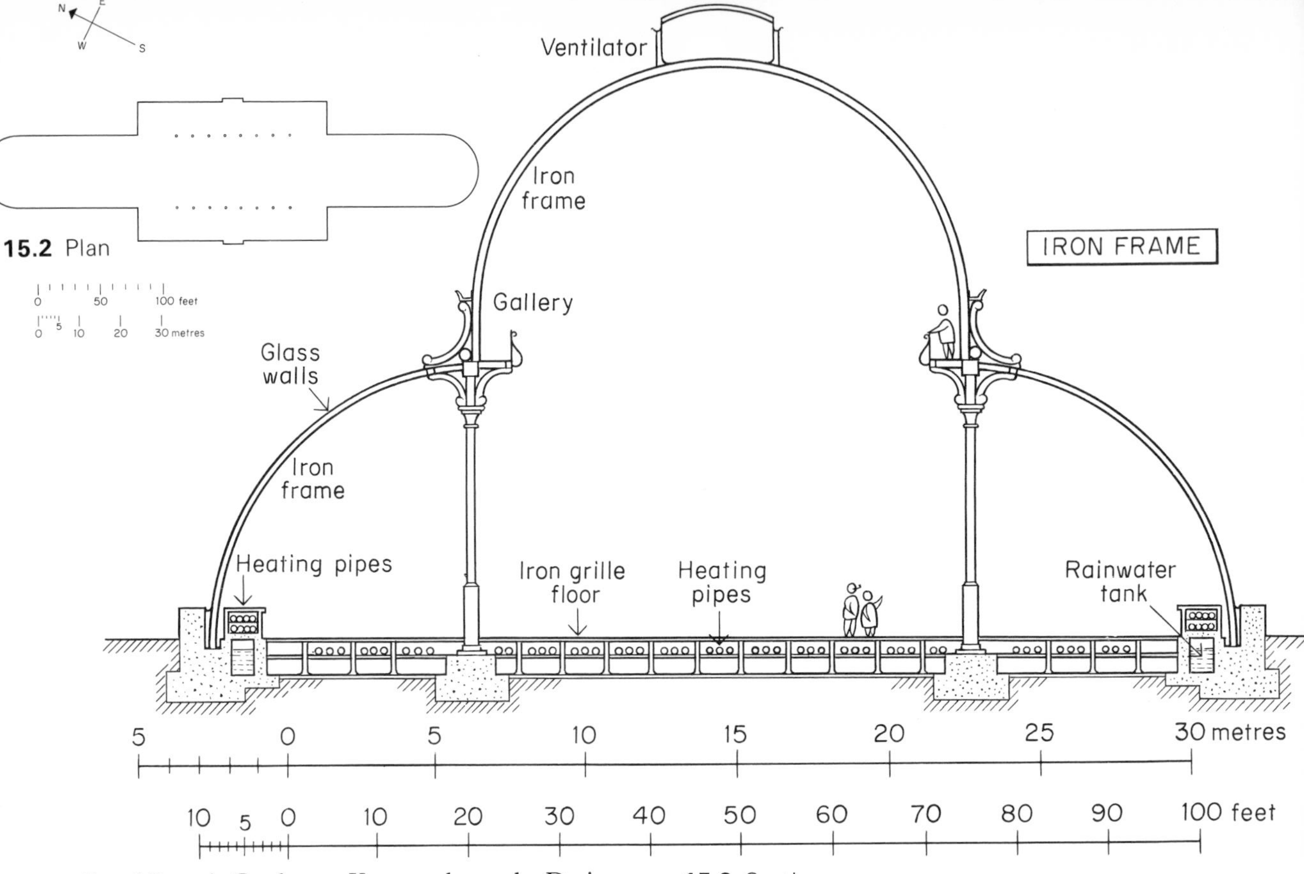

15.3 Section

Royal Botanic Gardens at Kew was begun by Decimus Burton in 1844. He asked Richard Turner of Dublin to work on the project with him and then to make the iron frames. Between them they produced one of the most beautiful of all iron-and-glass buildings which people today still look at with as much wonder as the Victorians did. (*Figure 15.1*)

Like earlier iron glass-houses Kew's glass walls swept up from the ground in double curves and met at the top 19 m (62 ft) above ground. And like Belfast's palm house two side wings joined a 32 m (106 ft) wide central hall. Heated air came up from under the floor and went out through windows at the top which could be opened or shut to control its flow. (*Figures 15.2 and 15.3.*)

Those Victorian families who were not overcome by the steamy air could climb the iron spiral stair to a gallery. From there they could look down on the strange leaves of the palms or look at the slenderness of Richard Turner's iron frames. These had been tied by wrought iron rods and cast as 203 mm (8 in.) deep **I** sections with an outer flange 100 mm (4 in.) wide and an inner flange 50 mm (2 in.) wide. (*Figure 15.4.*)

When the families came out into the open air of Kew again after their brief stay in a tropical climate they saw how the rain streamed down the curving walls of glass and was taken at the bottom by pipes into tanks below the floor.

Nobody could have guessed in 1848 that a 563 m (1848 ft) long Crystal Palace would go up in central London's Hyde Park during 1850 to the designs of Joseph Paxton. The Crystal Palace was to be ready for the summer opening of the 1851 Great Exhibition which was sponsored by Queen Victoria herself and her husband Prince Albert.

The iron-and-glass Crystal Palace opened on time because Joseph Paxton had the help of William Barlow who designed tracks and bridges for the Midland Railway. Barlow worked on the logistics for the thousands of cast and wrought iron parts that went into the Crystal Palace. Eight different foundries were casting and forging the parts and Barlow organised the vast process of manufacture so that iron beams and columns arrived in an orderly and non-stop flow of delivery at Hyde Park.

This experience was to help Barlow in 1866 when he built an even bigger iron-and-glass structure at St Pancras in central London. The district was one of London's worst slums. Cholera and other diseases killed hundreds of people in St Pancras. In 1866 the slums were demolished and the people made homeless as the site was cleared for the next marvel of the iron-and-glass age which still exists today as St Pancras Station.

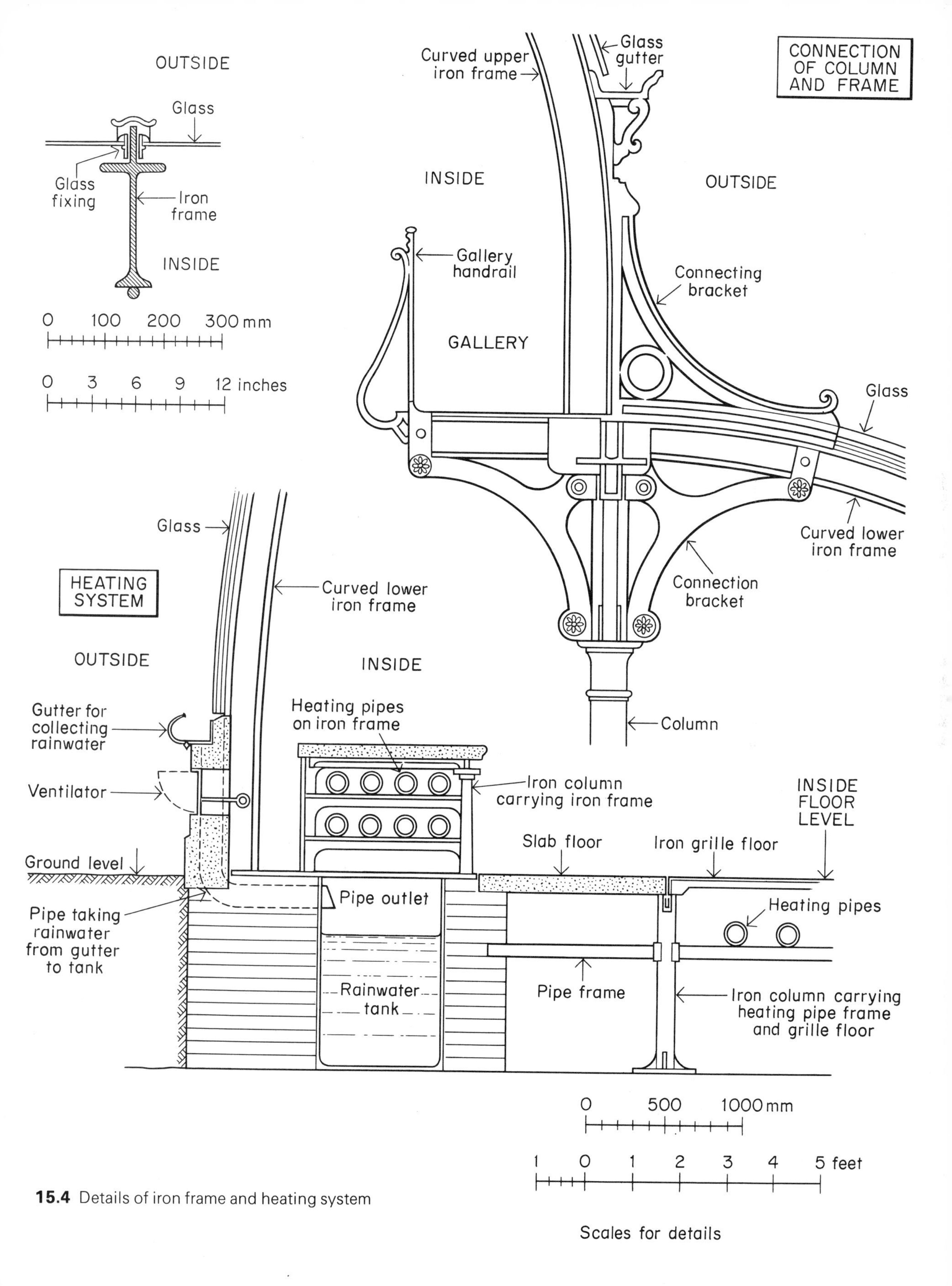

15.4 Details of iron frame and heating system

16 St Pancras Station

Place St Pancras, London
Time *Start* 1866
Finish 1868
Construction time $2\frac{1}{2}$ years
Purpose A main line railway station
Style Victorian industrial

16.1 The great scaffold. A photo taken in 1867 by the works foreman

The glass dome of the Anthaeum at Hove was not the only iron structure to fall down. A bridge at Stanford designed by the famous John Nash collapsed in 1795 and in 1803 so did a bridge over the River Thames at Staines. Even the great bridge-builder, Thomas Telford, could make mistakes. He designed a suspension bridge for Runcorn with a 305 m (1000 ft) span which luckily was not built; as mathematics have since shown, it would have collapsed.

The biggest crash of all occurred in 1847 when the bridge over Scotland's River Tay collapsed, taking trains and passengers with it. This disaster showed that much more work was needed on structural mechanics and strength of materials for large spans. It also showed that cast iron was a bad material for large spans with heavy moving loads like trains. Designers had already begun to use wrought iron as well as cast iron in beams and because wrought iron was strong in tension they put the metal in the tension zones of **I** beams and used cast iron for the compression zones. In 1847 when the Tay bridge crashed Henry Fielder got a patent for a cast iron **I** beam with wrought iron plates rivetted on to give extra strength. By 1850 no designers were using cast iron for large spans. This material was no longer considered safe, so wrought iron took over. (*Figure 16.5.*)

Wrought iron had been used for the first time in 1807 in a bridge at Boston in Lincolnshire. By 1830 railway tracks were being made of rolled wrought iron. In 1841 the first wrought iron girder was used for the 9·6 m (31·5 ft) span of a bridge near Glasgow.

Though John Nash was one of the first designers in the Classical style to put structural cast iron columns in his buildings, he was also among the last who did so without special training in structural mechanics. The ability to build with iron needed special training and as a result a special kind of designer emerged, the **structural engineer** or **civil engineer**. This new kind of designer used his mathematics and his knowledge of materials for large structures like bridges as well as smaller structures like the iron framing for buildings. The overall design and layout of a building still went on as before but it too became a specialist job and was done by the **architect**.

As time passed the job of each type of designer became so specialised that they understood less and less of each other's work and so the design of buildings was split between two professions. The knowledge, training, methods of work and, in the end, even the professional institutes for engineers and architects became quite separate. The Institution of Civil Engineers began in 1818 and the Royal Institute of British Architects in 1834. Nor did the split in the professions stop there: the municipal engineers set up their own organisation in 1873, the heating and ventilating engineers in 1897 and the structural engineers also became a separate profession in 1908.

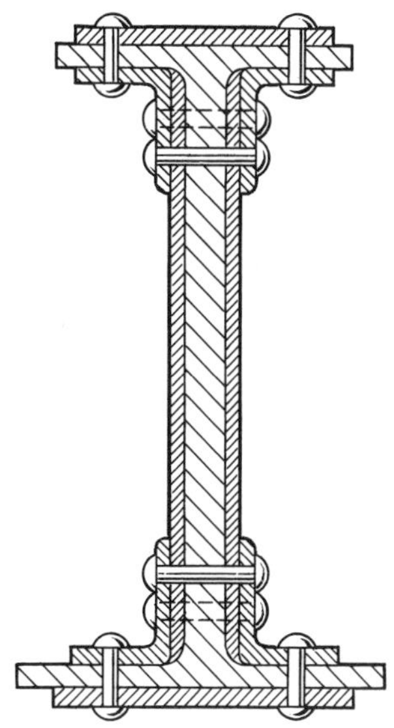

16.5 Henry Fielder's patent iron beam

The whole building industry in Britain slowly began to take the shape it has today when these various professional institutes were joined in 1865 by a contractors' organisation called the General Builders' Association which in 1878 became the National Federation of Building Trades Employers. Building workers had long ago started to form trade unions such as the 1827 Union of Carpenters and Bricklayers. The customers had their organisations too, for ever since 1775 they had been able to borrow money from Building Societies.

Many of the engineers who designed wrought iron bridges for trains to run over also designed wrought iron stations for the trains to stop at. Among the cleverest of these railway engineers was William Barlow. He had helped Joseph Paxton with the 1851 Crystal Palace and in 1863 finished the famous Clifton suspension bridge over the Avon Gorge at Bristol which the equally famous engineer Isambard Kingdom Brunel had begun in 1830. But Barlow was to be best known for the roof over St Pancras Station in London which had the widest span in the world. (*Figures 16.1 to 16.4.*)

St Pancras Station's 73 m (240 ft) span kept this world record for 21 years until the engineer Victor Contamin built a span of 114 m (374 ft) for the Machine Gallery at the Paris Exhibition of 1889. For the same exhibition another great French engineer Gustav Eiffel used wrought iron for the 300 m (1000 ft) high Eiffel Tower, a structure so vast that Wren's St Paul's Cathedral would fit with head-room to spare under just the lower part of it.

16.2 Looking down on the platforms

William Barlow lived from 1812 to 1902 and he was the ideal professional engineer of his time. His father taught mathematics and wrote a book about the strength of materials, and in 1855 William himself wrote about the strength of beams. The Royal Society printed some of his ideas and made him a Fellow as Christopher Wren and William Strutt had been before. Unlike many engineers Barlow thought that research was a vital activity and from 1858 he worked hard to bring steel into use for engineering structures.

The first railway stations were built in 1830. Of the many hundreds most were small and put up by country craftsmen using traditional methods and materials. But many larger stations were also built and the main part of them was the train shed which was a long roof over the tracks and platforms. Between 1830 and 1868 spans of these roofs grew from 10·7 m (35 ft) to 73 m (240 ft). Wood was the main material used for the early train sheds and wrought iron for the later ones.

The **rolling** process for wrought iron was used to produce ready-made structural members that just needed to be fixed together on site. The foundry made the sections by heating the wrought iron again until it was soft enough to pass between shaped rollers. The lengths of iron squeezed out then had the shape given by the rollers. By this method **sections** of wrought iron could be rolled quickly in great quantities. Rods and bars of wrought iron were already being rolled by 1700 and **L** sections by the 1820s. **T** sections were rolled by 1830 and **I** sections in 1847 when a Paris engineer Ferdinand Zores got a foundry to try out his idea. **L, T,** and **I** sections are still used today though now they are rolled in steel.

Large beams could be rolled in 12 m (40 ft) lengths by 1863 with depths up to 900 mm (2 ft) and with 300 mm (1 ft) wide flanges. But it was cheaper for large beams to be built up from plates and L sections. Big wrought iron plates were being rolled by 1863 in 5 m (16·5 ft) lengths which were 2·5 m (8 ft) wide and 112 mm (4·5 in.) thick.

William Barlow could use any of these ready-made wrought iron materials for his 1866 roof at St Pancras Station and for its design he could study other large train sheds that had been put up during the previous 30 years. Some of the main sheds were:

1 Crown Street, Liverpool
Built in 1830 with wooden trusses spanning 10·7 m (35 ft). This was the first of the large train sheds.

2 Temple Meads, Bristol
Built in 1840 with wooden hammer beams spanning 22 m (72 ft) was only 1 m (3 ft) wider than Hugh Herland's hammer beam roof built at Westminster Hall 450 years before.

3 King's Cross, London
Built in 1851 with arches laminated in eight layers of wood and spanning 32 m (105 ft). The 80 arches slotted into cast iron **shoes** at platform level and rose to a height of 22 m (72 ft). Steel arches replaced them in 1869 after the bad effect of steam and smoke on the wood.

4 Euston, London
Built in 1835 with triangular trusses of wrought iron rolled sections and bars. This was the first roof in which rolled sections were used in compression zones. The trusses rested on cast iron columns.

5 Lime Street, Liverpool
Built in 1849 by Richard Turner of Dublin who had just finished the Palm House at Kew. The engineer William Fairbairn worked out the structural mechanics of the roof which spanned 47 m (154 ft) with lightweight trusses. The top member was an arc of wrought iron L sections and plates made stiff by wrought iron tie bars and six struts.

This roof was one of the first **tension structures** and was held up by thin rolled metal in tension rather than thick cast metal in compression. Even though the amount of metal was small for such a span each truss was built-up with more than 41 tonnes of wrought iron. To cover his tension structure Richard Turner used a new type of sheet material. This was thin sheets of wrought iron rolled with a wave-like shape. For protection against rust the sheet was galvanized. This new product was **corrugated iron** which was soon exported in great quantities and which has been made ever since. Another new material at Lime Street was the rolled plate glass in the roof. The sheets were laid in three 3·8 m (12·5 ft) wide rows down the roof's whole length.

6 Newcastle Central, Newcastle
Also built in 1849 and designed by John Dobson who was an architect as well as an engineer. He had invented a way of rolling **L** and **T** sections so that they were curved as segments of circles. This was done with bevelled rollers. The segments could then be fixed end-to-end in a built-up **I** section beam curved like an arch.

That sort of arched beam needed no tie rods to stop its ends from spreading outwards because the **I** section was itself stiff enough. Dobson used these curved **I** sections for the 16·7 m (55 ft) spans at Newcastle Central. Many other train sheds were built by the same method later.

7 New Street, Birmingham
Built in 1854 with one immense span of 65 m (211 ft). The engineer John Hawkshaw helped in its design as he was a leading expert on wrought iron structures and worked later with William Barlow on the completion of Brunel's Clifton suspension bridge at Bristol.

8 Cannon Street, London
Built in 1864 as another roof with a single span, this time of 58 m (190 ft) also designed by John Hawkshaw. Single-span train sheds were good because if trains came off the rails there were no pillars carrying the roof for them to run into. There had already been one accident which brought a roof down. William Barlow studied the design of Cannon Street's roof and he talked to Hawkshaw about it. It was the roof design which helped Barlow to make up his mind to use a huge single span at St Pancras Station.

From his training as an engineer and from experience such as his work at the 1851 Crystal Palace, William Barlow had learnt a method of design which today is called **problem solving**. By this method he studied a project and sorted it into a number of problems for which solutions had to be found. Each problem was distinct but linked to all the others. Barlow soon found that he had to solve two main problems at St Pancras: the tracks and the roof. (*Figure 16.4.*)

Problem 1: The tracks
Barlow had an unusual problem to solve because the tracks came from the hills and new tunnels north of London into the station, 4·5 m (15 ft) above street level. It was not possible to slope the tracks down to street level because the engines pulling trains in 1864 were not powerful enough to climb up such steep gradients. So Barlow had to solve the problem of keeping the tracks and their heavy moving loads 4·5 m (15 ft) above street level. He had two ideas.

(*i*) *Fill* A thick wall could be built around the site which could then be filled up to the 4·5 m (15 ft) level with the soil that was being dug out for the tunnel linking St Pancras Station with the underground Metropolitan Railway. This link was to give goods and coal trains coming into St Pancras from Britain's industrial Midlands access to Kent and the south-east region.

To support the tracks on filled ground would be quick and cheap. But it would also stop some profit-making street level space from being used. So Barlow gave up this idea.

(*ii*) *Floor* A floor could be built 4·5 m (15 ft) above street level for the incoming tracks. This floor would carry the tracks and also make vast storage space at street level. Barlow did not hesitate to choose this method.

Barlow built the tracks and platforms on ballast laid over wrought iron plates which were held up by

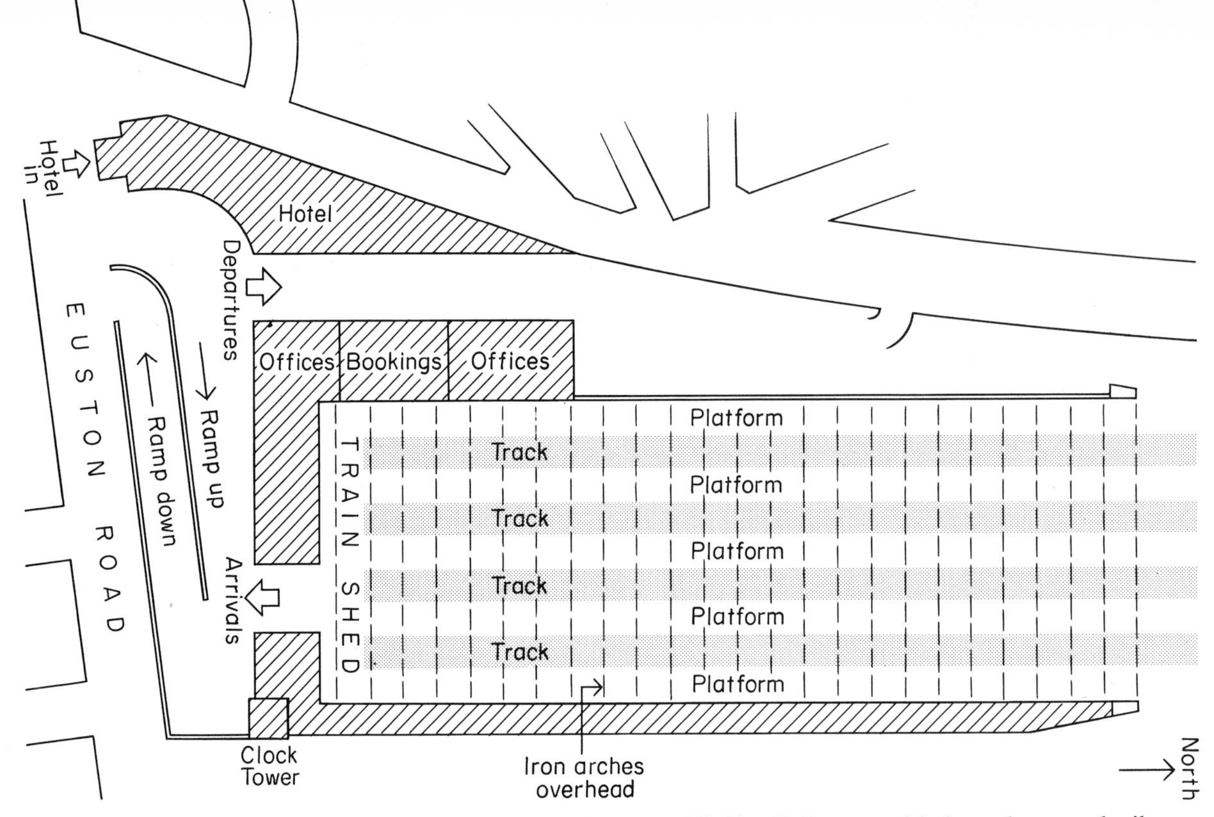

16.3 Plan

wrought iron girders and cast iron columns. Immense foundations of brick would take all the weight of the iron structure and the moving trains down to firm ground. This structure would make a vast warehouse with around 1·6 hectares (4 acres) of floor space at street level.

The cast iron columns were spaced the same distance apart as those in beer warehouses at Burton-on-Trent. Burton beer would come by train in great quantities to St Pancras Station so the street level warehouse would be ideal. Hydraulic lifts could take the beer down from platform level to warehouse level. Drays pulled by horses would be able to come straight from the street outside into the warehouse for their barrels.

Barlow then had to solve the second main problem of his design.

Problem 2: The roof

The total width of tracks and platforms would be 73 m (240 ft) and there were two methods of putting a roof over them: multiple spans and single span.

(*i*) *Multiple spans* The total width of the roof could be divided into a number of smaller spans as Brunel had done at London's Paddington Station in 1852. But such spans would have to be held up on iron columns and columns took up space and needed deep foundations especially here at St Pancras because they would also have to carry the floor beams with the weight of trains on top. Some of the foundations would be on top of the tunnel linking St Pancras with the underground railway and so the tunnel would need to be strengthened.

Columns would have to be at least 18 m (60 ft) long and would need girders at the top between one column and the next to ensure stability. Such a scheme would lead to huge costs with no extra benefit in return, so Barlow gave up the idea of multiple spans.

(*ii*) *Single span* Barlow had studied the single span roof over Cannon Street Station which his friend John Hawkshaw had designed and built in 1864. It was a truss with two main members. The first was the top **chord**, a wrought iron I section curving up like an arch across the whole 58 m (190 ft) space. The second was the tie of wrought iron bars fixed between the bottom ends of the chord to stop them spreading outwards. There were also struts to stiffen the structure.

The distance from the top of the arched chord to the bottom tie was called the **effective depth** of the truss and was usually 0·2 × span. So a truss spanning 58 m (190 ft) would have an effective depth of 11·6 m (38 ft). This would be the height that the top chord would arch upwards from the supporting walls.

At St Pancras Barlow decided to use a truss with an arched top chord and a bottom tie. He knew the span had to be 73 m (240 ft) but he had a clever idea for making the arch higher and at the same time using less material. Instead of the usual effective depth of 0·2 × span, Barlow doubled it to 0·4. This gave 29·2 m (96 ft) as the height to the top of the truss. Through his knowledge of the theory of structures he worked out that the wrought iron parts for the truss could be sized as if the truss only spanned 36·5 m (120 ft) with an effective depth of 0·2 × span. This would reduce the amount and weight of iron in each truss. (*Figure 16.4.*)

Barlow's next idea was clever and simple. The single span truss at St Pancras was going to be 29·2 m (96 ft) high. This was more than high enough for all the platforms and tracks to fit underneath the arch. The arch could therefore start its rise from platform level and its tie could go under the platforms. But because the platforms and tracks had to be held up by girders Barlow could use those girders as the tie position. This clever solution to the two problems produced one of the most striking of all Victorian iron-and-glass structures.

There were two main contracts at St Pancras: the station and the roof.

Contract 1: The station

This work was for the London end of the Midland Railway's new line coming down through Bedfordshire to tunnels under the hills of Hampstead to the north of London and so down to St Pancras. In December 1865 this contract was let to Waring Brothers for the construction of the railway between St Pancras and a point just north of Regent's Canal where it would join the rest of the new line being built by other contractors. This distance of nearly 1220 m (4000 ft) included the tracks themselves with their cuttings and embankments and four new girder bridges as well as the 198 m (650 ft) long station and its tunnel down to the underground Metropolitan Railway. Waring Brothers' estimate, or **tender price**, for this work was £319,000. Barlow had expected it to be £310,000. A civil engineering contract of this size would be tens of millions of pounds today.

Progress on site for the station went as follows:

1866 In the spring Waring Brothers moved onto the site and in June cleared all the old slums away and began the new foundations in July. There were 688 brick piers to be built as foundations for the cast iron columns supporting the platforms and tracks 4·5 m

16.4 Cross-section and diagrams of Barlow's idea

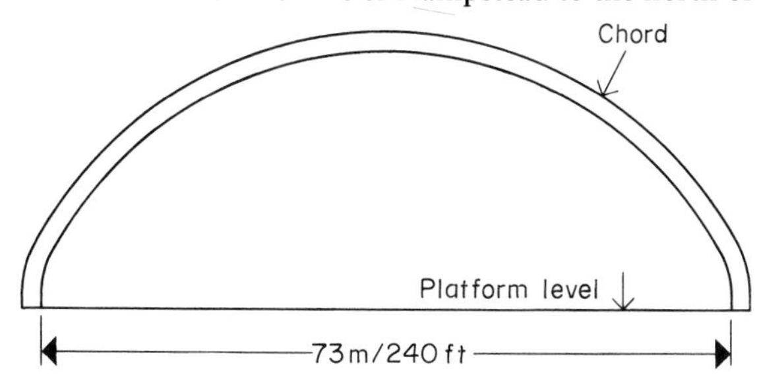

1 SINGLE SPAN
Rising from platform level

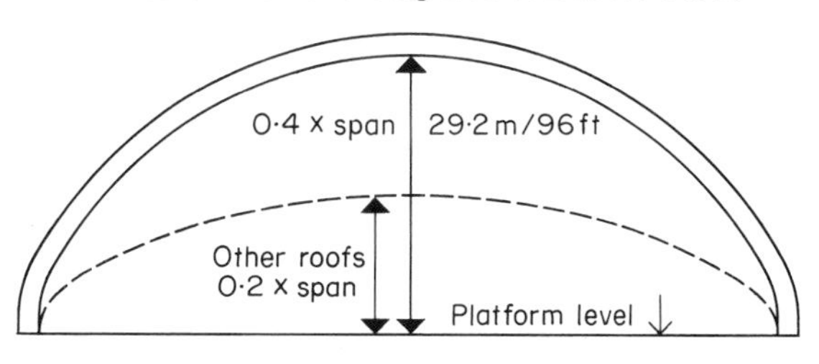

2 DOUBLE DEPTH
$0{\cdot}4 \times \text{span} = 0{\cdot}4 \times 73\,\text{m} = 29{\cdot}2\,\text{m}$ (96 ft)

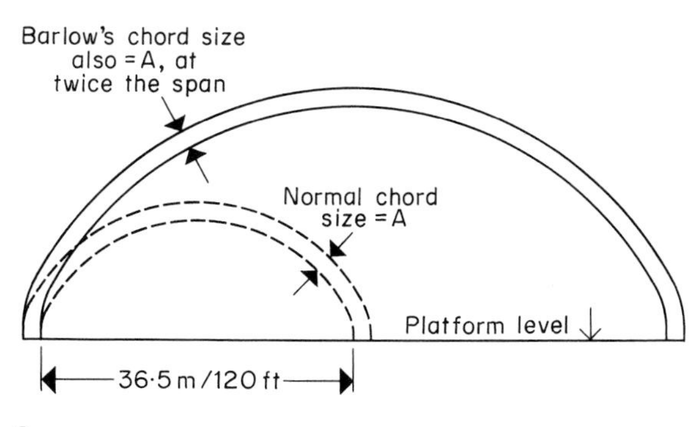

3 SLENDER CHORD
Double depth allows chord to be same as a chord for only a 36·5 m (120 ft) span

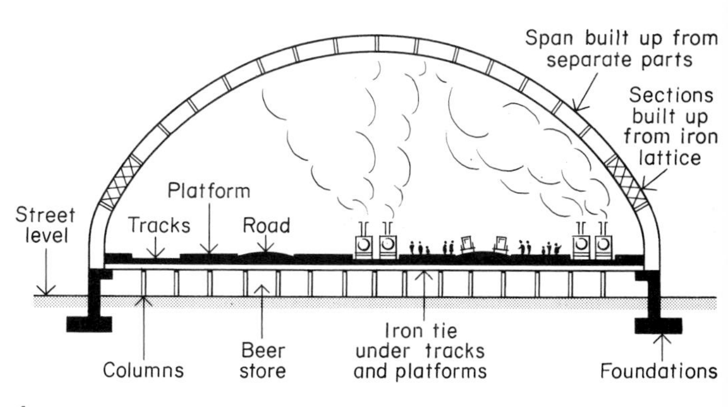

4 COMPLETE ROOF

(15 ft) above street level. By October the contractors had 988 men on site with 47 horses, 18 steam engines and one locomotive.

1867 In January the nearby Fleet River had to be made into a proper sewer and was enclosed in a large iron pipe. The river had long been one of London's worst open sewers and partly caused the high level of diseases, such as cholera, in St Pancras. Even while Waring Brothers were working on the site a cholera epidemic raged in the district all around.

By May six bases or **springers** for the great wrought iron arches of Barlow's roof were in place. And in the autumn Waring Brothers were also given the contract to build the long side walls of the train shed. The contractors were to use Gripper's bricks.

Edward Gripper was an Essex farmer who had moved to Nottingham in the 1850s and in 1867 set up a company with a large brickworks there for making patent bricks. Gripper had a licence which allowed him, but nobody else in the district, to use Hoffman's continuous burning kilns invented in Germany in 1856. By this method bricks were loaded, burnt and cooled in a ring of chambers with the fire passing non-stop from one chamber to the next. This was faster than the traditional method where the fire had to be started and then left to burn itself out for each batch of bricks. But even by this rapid method Gripper could not make the bricks fast enough and progress was held up on site.

Gripper's prices varied according to the season of delivery. From November to April **best fronts** were 50 shillings per 1000, 'cappers' 37 shillings-and-sixpence and **commons** 26 shillings-and-sixpence. Summer prices were slightly cheaper. Another brickworks called Gray's also tendered with 36 shillings for the best and 21 shillings for commons all year round. Today the same sort of bricks would cost many times as much.

Before the end of 1867 all the foundations were finished.

1868 In July the tunnel down to the Metropolitan Railway was finished.

Contract 2: The roof The contract for the vast wrought iron roof was let in July 1866 to Butterley and Company who had their ironworks in Derby and who had other contracts with the Midland Railway. Butterley's tender price was £117,000. Barlow had allowed £121,000.

The roof was to be built by putting up the first of the wrought iron arches at the south end of the train shed where Euston Road is today and then working in sequence northwards until the last of the 25 vast arches was in position.

An immense mobile scaffold on specially laid railway track was built with baulks of timber. It acted as centering for the iron arches and was built to the same shape, 73 m (240 ft) wide at the bottom and 29·2 m (96 ft) high. It was 28 m (90 ft) deep and made in three separate parts. Teams of navvies moved the vast scaffold from one arch to the next by heaving on levers between the wheel spokes. The 1300 tonnes of the scaffold took two hours to travel between the arches.

The separate wrought iron components of each arch were hoisted up and laid on the centring and fixed together to make the huge curving chord. When each arch was finally self-supporting the heavy scaffold was pulled along the rails into position for the next arch to be built.

Progress on site for the roof was made as follows:

1867 The first arch was begun in November. This was one year later than the date agreed in the contract. Butterley had failed to understand the vast extent of the project and they could not roll the wrought iron parts fast enough.

1868 By February the construction of the arches at last got under way smoothly and quickly. Four were already in position and in March a second mobile scaffold was built so that three or four arches a month went up. By July the first glass and slates were going onto the roof and in September the last of the 25 arches was in place. The first mobile scaffold was dismantled and its timber sawn up to make wood paving for the station platforms. On 1 October 1868 the first passenger train ever to leave St Pancras Station pulled out and ran non-stop to Leicester in a record time of 134 minutes. It had taken over two years and £436,000 to get that train moving.

There were other big contracts at St Pancras for the railway offices and hotel building designed in the Gothic Revival style by the architect George Gilbert Scott. This was sited at the Euston Road end of the train shed and the contractors, Waring Brothers, were already working on the hotel's foundations in March 1867 when the contract for the upper part of the building was let to Messrs Jackson and Shaw. The hotel opened for guests in May six years later but it was 1876 before Gilbert Scott worked out the final account of the hotel's cost at £483,000.

The shareholders in the Midland Railway Company had to wait until 1877 before being told that the total project of new lines and station and hotel had cost nearly £1 million, a figure that would be many millions more today. But by 1877 the hotel's spiky Gothic Revival skyline and the station's cavern of wrought iron had become well-known landmarks in mid-Victorian London. St Pancras Station was not only one of the most daring of all iron-and-glass structures but also it was one of the last. In 1883 the engineer Benjamin Baker began the Firth of Forth bridge in Scotland. Its cantilevered spans of 500 m (1640 ft) were built in steel. The rule of iron was over.

PART 5 **HIGH TECH**

17 Royal Corinthian Yacht Club

Place Burnham-on-Crouch, Essex
Time *Start* and *finish* 1931
Construction time 1 year
Purpose Club rooms and guest rooms
Style The International Style. Also known as the Functional Style or the Modern Style

The Royal Corinthian Yacht Club built at Burnham-on-Crouch by the English architect Joseph Emberton is a perfect example of the International Style. This style began in the 1920s and ended in the 1930s but though it had a short life it had a long-lasting effect on designers. The style was marked by its almost total break with styles of the past and its use of reinforced concrete as a structural material.

17 Royal Corinthian Yacht Club, Burnham-on-Crouch, Essex, 1931 *Designer* Joseph Emberton

In the 1850s designers were trying out new styles that were not Classical or Gothic Revival and by the 1920s a number of quite different styles developed. There were four main groups:

Group 1: New styles from new materials
Designers in this group used the most up-to-date materials and methods of construction. There were many examples in Britain, Europe and America. Some of the best known were:

Reliance Building, Chicago, USA Built in 1890 this 16-storey office block had one of the first multi-storey steel frames. Its designer was Daniel Burnham who later did the basic layout and construction of Selfridge's in London. (*Figures 17.1 and 17.2.*)

Flats at 25 Rue Franklin, Paris Built in 1902 this block had one of the first multi-storey frames in reinforced concrete, or **rc** as the material was often called for short. The engineer and architect Auguste Perret constructed the building by rc methods he invented and which were later used all over the world. (*Figures 17.3 and 17.4.*)

Ingall's Buildings, Cincinnati, USA Built in 1902 this 16-storey office block was one of the first skyscrapers with an rc frame. The architects of this 65 m (210 ft) tall building were Elzner and Anderson. (*Figure 17.5.*)

Fagus factory, Alfeld, Germany Built in 1911 by the architect Walter Gropius. The glass and steel panels outside went between the building's brick piers like a wrapping and at corners all the structural uprights were placed away from the outer edge of the floor. This **cantilever** of floors leaving their edges free of vertical support was to become a main feature of the International style. (*Figure 17.6.*)

St Thomas's School, Birmingham Built in 1914 by the architects Harrison and Cox, this was one of Britain's first buildings with rc used as an outside as well as an inside material. Like many thousands of buildings since, some of the rooms were raised above the ground on rc uprights. The flat rc roof was a playground.

17.1 Reliance Building, Chicago, 1890
Designer Daniel Burnham

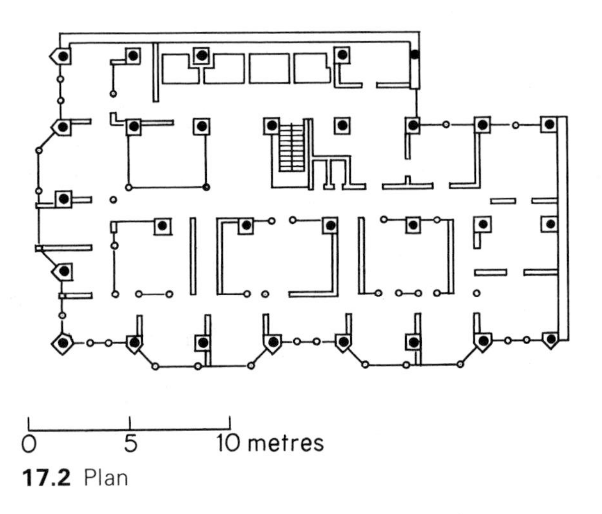

17.2 Plan

17.3 Flats at 25 rue Franklin, Paris, 1902
Designer Auguste Perret

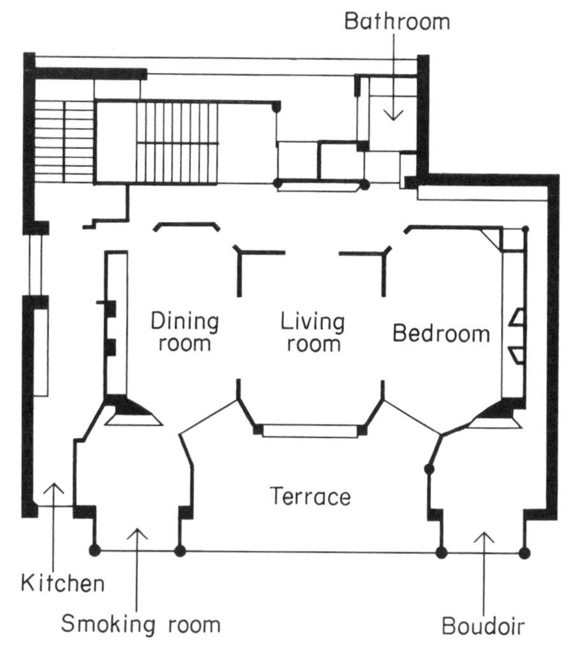

17.4 Plan, rue Franklin

17.5 Ingall's Building Cincinnatti, 1902
Designers Elzner and Anderson

17.6 Fagus Factory, Alfeld, 1911
Designer Walter Gropius

17.7 The Red House, Bexleyheath, 1859
Designer Philip Webb

Group 2: New styles from old crafts

Designers in this group got their ideas from country buildings. They used old craft methods and materials. These ideas led to the 'Arts and Crafts' group after 1884. There were many examples in Britain. Among the best-known were:

The Red House, Bexleyheath, near London Built in 1859 by the architect Philip Webb for his friend William Morris. This was the first house built in a plain country style with red brick and tile. It looked like a big farmhouse that had been added to over the years. Many thousands of much smaller houses were later to copy the Red House ideas. (*Figures 17.7 and 17.8.*)

The Cottage, Bishop's Itchington, Warwickshire Built in 1888 by Charles Voysey, the most well-known of the 'Arts and Crafts' architects. It was not a cottage at all but a big house. The word 'cottage' helped the country effect aimed at. Voysey used white walls with rough surfaces and long bands of windows. (*Figure 17.9.*)

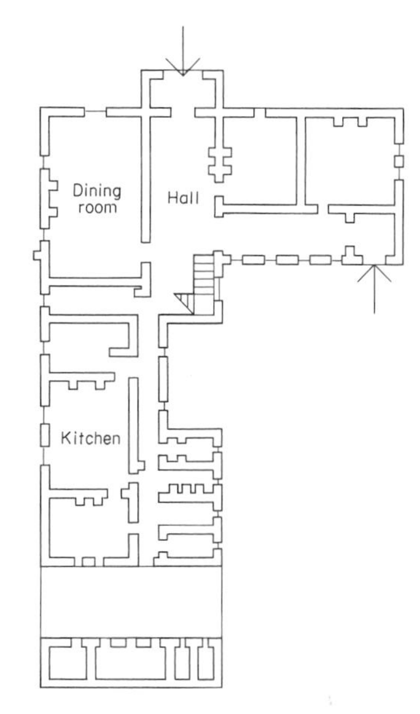

17.8 Plan, The Red House

17.9 The Cottage, Bishops Itchington, Warwickshire, 1888 *Designer* Charles Annesley Voysey

Group 3: New styles from nature
Designers in this group wanted buildings to have links with nature and be as good for their purpose as the structure and shapes of plants and animals were. In 1853 the German designer M H Riehl said that buildings should get their shapes from the needs of the people using them and not from axis lines or grids of rectangles as in the Classical style. And in 1856 the French writer César Daly said that decoration should come from local plant life and not from Mediterranean plants such as the acanthus used by the ancient Greeks and Romans.

In 1863 Daly also said that buildings of the future should grow out of up-to-date needs in the way that branches grow from a tree trunk. Design should be organic as nature was organic. These ideas were basic to the International style of the 1920s and 1930s.

By the 1880s the shapes and curves of stems and leaves had become one of the main new styles called **Art Nouveau** – the New Art – after a shop of that name that was opened in Paris in 1883. People liked the strange and wavy shapes of objects the shop sold and the style spread all over Europe. But designers found the style more suited to furniture and small objects in precious stones and metals than to buildings. But some well-known buildings were designed in the Art Nouveau style. Among them were:

The Tassel House, Brussels Built in 1892 by the Belgian architect Victor Horta. The decoration of the entrance hall and the handmade wrought iron of the staircase were given the curling lines of glass-house plants. (*Figure 17.10.*)

17.10 Tassel House, Brussels, 1892
Designer Victor Horta

17.11 School of Art, Glasgow, 1896
Designer Charles Rennie Mackintosh

The School of Art, Glasgow Built in 1896 by the Scottish architect and furniture designer Charles Rennie Mackintosh. Like his other work this building was far less leafy-looking than Art Nouveau buildings in Europe but its ironwork railings and decorations and furniture had a hint of long plant stems. (*Figures 17.11, 17.12 and 17.13.*)

Metro entrances, Paris Built around 1900 by the architect Hector Guimard who designed three standard street-level entrances for the underground stations of the Paris Metro. They all had writhing iron and glass signs and roofs and gates at street level in the Art Nouveau style. (*Figures 17.14 and 17.15.*)

Flats at Avenue Rapp, Paris Built in 1901 by the architect Jules Lavirotte. This block used a new material invented by Alexandre Bigot who had found a method of giving fireclay blocks and tiles a surface glaze which could be coloured and was hard like enamel. The blocks could be moulded to the architect's design and held in position by steel rods passing through specially made holes in the blocks. So the result could be a highly coloured, wavy surface for the outside of a building like the surfaces of smaller Art Nouveau objects. (*Figure 17.16.*)

Architects used any material or method, old or new, yet they soon knew that the Art Nouveau style could never be used for the basic structure of a building but only to decorate inside and outside surfaces. The Spanish architect Antoni Gaudi had tried to give Art Nouveau curls and swirls to basic construction. The results, still to be seen at Barcelona in such buildings as the 1905 Casa Mila block of flats, were striking but no good for everyday use. (*Figure 17.17.*) Another example is Gaudi's extension to the Holy Family church in Barcelona with its curving spires carried out between 1884 and 1926. (*Figure 17.18.*) The Art Nouveau style was a highly personal style and it cost too much. By 1910 few designers were using it.

17.12 Detail of ironwork
17.13 Plan

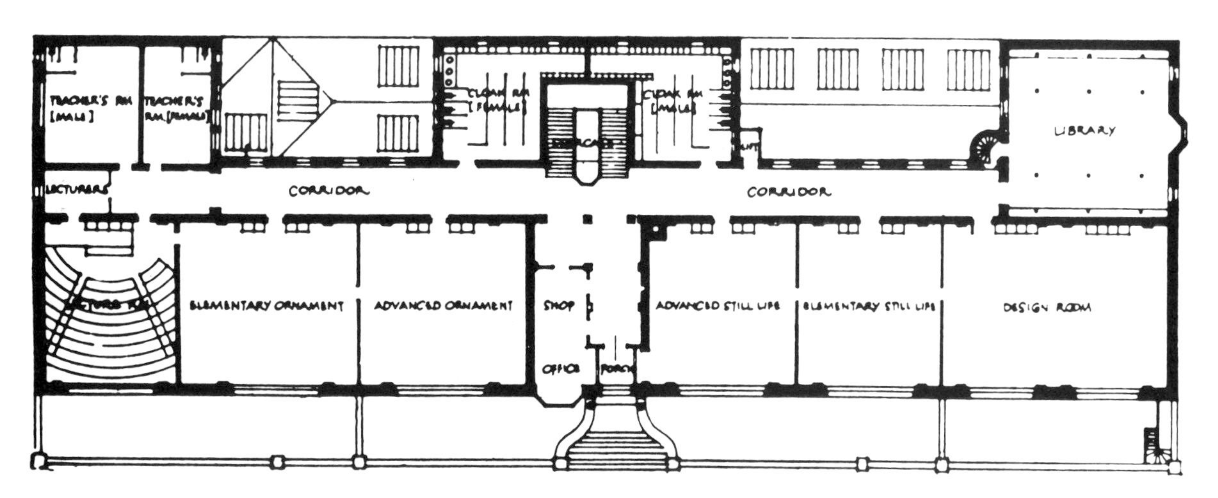

17.14 Metro entrance, Paris, *c.* 1900
Designer Hector Guimard

17.15 Sketch

17.16 Flats at Avenue Rapp, Paris, 1901
Designer Jules Lavirotte

17.17 Flats at Casa Mila, Barcelona, 1905
Designer Antoni Gaudi
17.18 Section of Church of the Holy Family, Barcelona
Designer Antoni Gaudi

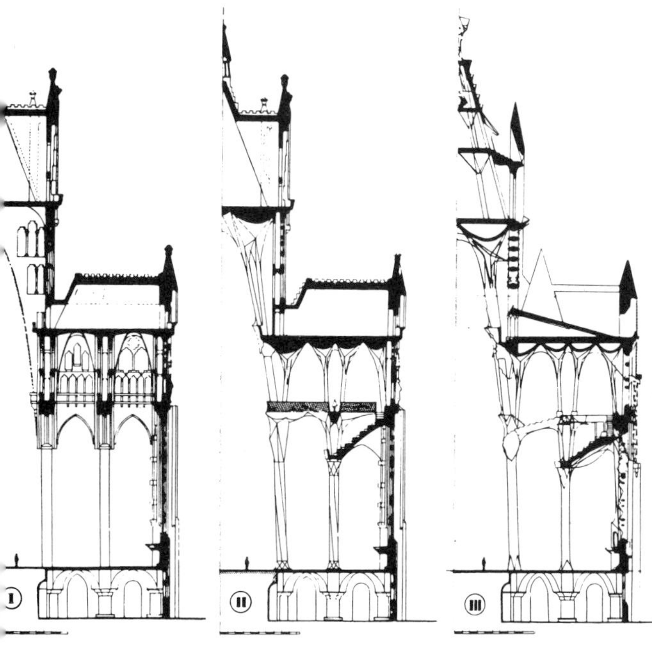

Group 4: New styles from new artists

Designers working in this style got ideas from several groups of new artists. Ideas from painting and three-dimensional art helped to invent the International style: up-to-date artists and designers in the 1920s wanted styles that in no way looked like any art or design from the past. Some of the main groups of new artists were:

Cubist artists The International style designers wanted to create buildings where outside spaces and inside spaces could be seen at the same time. They made stairs hang in mid-air so that people going up or down them could see right through the building and through its glass walls, and so get changing views of it.

The designers saw ideas like these in the paintings of Cubist artists such as Braque and Picasso. Like other artists the Cubists painted everyday objects referred to as still-life pictures. But Cubist still life showed many views of the object all in the same picture. By opening up buildings with glass walls and mid-air stairs International style designers could show many views of the building all at once.

Abstract artists The International style designers also wanted their buildings to have the simple shapes of geometry such as squares and circles, cubes and cylinders. Designers saw these in the abstract paintings of artists such as Piet Mondrian and Theo van Doesburg who made three-dimensional abstract sculptures with rectangular prisms. The work by this group of new artists was called 'Abstract' because the paintings were not recognisable as pictures of people or objects; they were brightly coloured geometric patterns.

Many architects in Holland took their ideas from Abstract artists. For example, in 1923 Gerrit Rietveld built the little Schroeder House at Utrecht. Its walls and windows and doors were all like the rectangular coloured planes of an Abstract sculpture. It did not look like a house at all. (*Figure 17.19.*)

Constructive artists The International style designers also wanted their buildings to look as if they were put together with up-to-date methods and materials such as reinforced concrete, steel and plate glass. They did not want buildings that looked as if craftsmen had built them with old methods and materials such as stone. The designers saw ideas like these in the work of new artists called 'Constructivists'. The ideas began in Russia after the 1917 Revolution and were brought to western Europe in the 1920s by some of the artists.

The buildings designed by International style architects were notable for the following features:

1 Frames The main structure for their buildings was a frame of columns and floor beams made of steel or rc with rc slabs for the floors and roof. Walls were not load-bearing and were only used as partitions, sometimes stopping short of the ceiling.

2 Cantilever Structural columns were kept back from the outer edge of floor slabs so that the floors had an overhang clear of vertical supports.

3 Glass Large sheets of glass were used as outside walls and long bands of windows ran across buildings in front of columns.

4 Ground floor Much, if not all, of the ground floor was left open as a void because the rest of the building was raised on steel or rc columns. This was done partly to destroy the Classical style's feature of heavy stone walls and rustication at ground level.

5 Flat roof This was often used as an outdoor room and as a roof garden. Construction in rc let the flat roof be any shape which gave freedom for the placing and shapes of rooms below.

6 Inside-outside The thin rc structure and large areas of glass made the inside and outside seem like part of the same space. Again this destroyed the Classical style's rigid separation by thick walls of the inside from the outside.

7 Double-height spaces Spaces often went up through two or more storeys. People could look over cantilever galleries into these double-height spaces. Cantilever stairs or ramps often went up in mid-air through them.

8 Plain and white walls All wall surfaces had to be flat and white and without surface decorations or mouldings of any kind, inside or outside. All materials that could be seen had to look as if they were machine made and machine finished, even when they were in fact hand finished. The Austrian architect Adolf Loos summed up these ideas with his slogan, *Ornament is crime*. In 1911 he built one of Europe's first rc houses, the Steinerhaus in Vienna. Its white cube shapes had no ornament at all.

Most countries in Europe had buildings in the International style. In France the architect Le Corbusier ruled. In 1922 he designed the Citrohan house and a studio for his artist friend Amédée Ozenfant. And in 1925 he designed the pavilion at the Paris Exhibition for the modern art group *L' Esprit Nouveau* – The New Spirit. Then he did the well-known 'white houses' near Paris, such as those at Garches in 1927 and at Poissy in 1929. (*Figures 17.20, 17.21 and 17.22.*) In 1931 he built a hostel for Swiss students at Paris University.

In Germany the architect Walter Gropius built the Bauhaus design school in 1925 at Dessau. (*Figure 17.23.*) Stuttgart got its Schocken department store in 1926 by the architect Erich Mendelsohn who did many other buildings such as the Berlin office, the 'Columbushaus', built in 1921–31. (*Figures 17.24 and 17.25.*)

17.19 Schroeder House, Utrecht, 1923
Designer Gerrit Rietveld

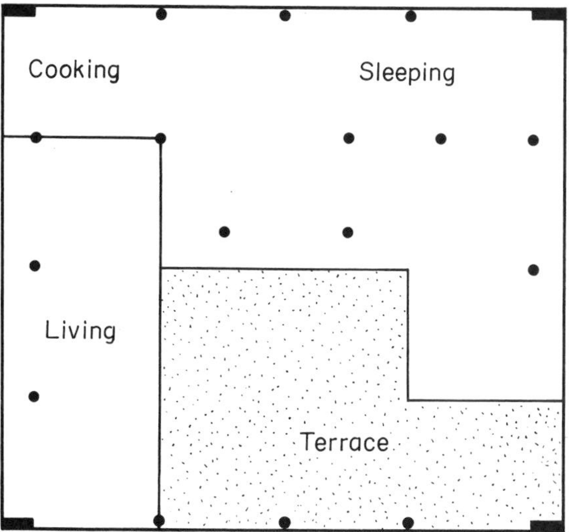

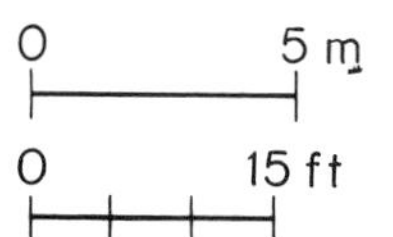

17.21 Plan

17.20 Villa Savoye, Poissy, nr Paris, 1929
Designers Le Corbusier and Pierre Jeanneret
When neglected, concrete buildings become uglier with age

17.22 Interior of Villa Savoye, 1929

17.23 The Bauhaus, Dessau, 1925
Designer Walter Gropius

17.24 Schocken Store, Stuttgart, 1926
Designer Erich Mendelsohn

17.25 Columbushaus, Berlin, 1921
Designer Erich Mendelsohn

In Holland the 1927 Van Nelle tobacco factory was built to the designs of Mart Stam. (*Figure 17.26.*) In 1928 Johannes Duiker designed Amsterdam's 'Open-air' school. And in Finland the architect Alvar Aalto built one of the International style's most striking examples, the hospital at Paimio between 1929 and 1933. (*Figure 17.27.*)

The work of new artists and designers and architects in Europe shocked most people in Britain who thought it all a new outrage rather than a new outlook. Yet some buildings went up in the International style.

New Ways, Northampton Built in 1926, this white, cube-shaped house was by the German architect Peter Behrens who was the chief designer for the German electrical firm AEG.

Silver End, Essex Built in 1927, these houses were for the Crittall metal window company and they were designed by Thomas Tait who was the young designer working behind the scenes on John Burnet's 1910 Kodak building. (*Figure 13.6.*)

High and Over, Buckinghamshire Built in 1929, this house was like Le Corbusier's 'white houses' and its architect Amyas Connell did similar ones elsewhere such as the group of three on the cliffs near Brighton. (*Figure 17.28.*)

Royal Corinthian Yacht Club, Essex Built in 1931, this was one of many buildings designed by Joseph Emberton, his 'Olympia' exhibition and arena building in London being known to millions of people. (*Figure 17* page 83.)

Boot's factory, Nottingham Built between 1930 and 1932, this rc structure supported the slab floors with 'mushroom' columns which the engineer-designer Owen Williams enclosed with glass walls. He was one of Britain's first experts on rc construction. (*Figure 17.29.*)

De la Warr Pavilion, Bexhill Built in 1935, this seaside leisure centre and theatre showed all the features of glass walls and horizontal lines for which its designer Erich Mendelsohn was well known in Germany. He was then in Britain as a political refugee. (*Figures 17.30 and 17.31.*)

17.26 Van Nelle tobacco factory, 1927
Designer Mart Stam

17.27 Paimio Hospital, 1929–33
Designer Alvar Aalto

17.28 High and Over, Old Amersham, Bucks, 1929
Designer Amyas Connell

17.29 Boots factory, Beeston, Nr Nottingham, 1930–32
Designer Owen Williams

17.30 De La Warr Pavilion, Bexhill, Sussex, 1935
Designer Erich Mendelsohn

17.31 Detail, De La Warre Pavilion

Before the Second World War started in 1939 Mendelsohn had gone to live and work in America where the International style was more welcome than in Britain. Walter Gropius also went to America after a brief stay in Britain where he designed the Impington Village College in Cambridgeshire in 1936 with the English architect Maxwell Fry who himself designed the white cube and cantilever 'Sun House' at Hampstead in London in the same year. (*Figure 17.32.*) Just round the corner in 1938 were to be the flat white walls and horizontal bands of windows designed for 'No. 66 Frognal' by the firm of architects Connell, Ward and Lucas.

The Second World War stopped nearly all building and when the war was over in 1945 the International style was over too. New methods and materials were needed to produce the thousands of homes and schools and all the other buildings either destroyed by the bombings or now so old that replacement was vital.

Yet the 20 years of the International style had shown architects that only a complete break with the Classical past could open the way to new ideas. In 1952 London began building its high-rise public housing project at Roehampton. It was one of Britain's first and best post-war schemes. The high-rise future looked good. (*Figure 18.1.*) Few people yet saw that this future would last no longer than the International style had done and end in social and environmental disaster.

Reinforced concrete was like cast iron in that it too was a material poured into moulds and had no shape until it was designed. And rc could not be designed until mathematics and structural mechanics had shown where bending stresses and other forces would occur inside the floors or columns or beams.

Still today the basic idea of rc is that the concrete takes the squashing or **compression** and the steel bars inside the concrete take the pulling or 'tension'. The steel bars are placed in the moulds and the concrete is poured around them and, in drying-out, grips the bars so tightly that concrete and steel act as one material. The earliest work on rc was done in Britain and France and later in Germany and the USA.

Britain In 1818 the engineer Ralph Dodd was given a patent for putting wrought iron bars in concrete, and in 1825 Thomas Telford put iron bars in the concrete abutments of the Menai Bridge. Fox and Barret received a patent in 1844 for their fireproof floor of cast iron beams buried in lime concrete which was used at Balmoral Castle and elsewhere. W B Wilkinson used wire ropes and iron bars in concrete beams and in 1865 he built a rc house that was still sound when demolished in 1954.

Many patents were taken out for various rc methods after 1850 and during the 1870s and 1880s the material was a main concern of engineers and architects who worked on the mathematics and mechanics of design and tried out new methods. But real progress was held back by a distrust of rc and rigid interpretation of building regulations by building inspectors.

17.32 'Sun House' Frognal Way, Hampstead, London, 1936
Designers Fry, Drew, Knight and Creamer

In 1894 the London Building Act took over from the old Metropolitan Building Regulations, but the new Act did not deal with any new methods of construction and the thin walls possible with rc had to be built as thick as brick walls. It was 1909 before another Act of Parliament let the London County Council, (LCC), make special building regulations for rc. But even as late as 1912 one designer was told by the LCC to make his rc walls as thick as brick ones.

By 1910 rc had become a common material for building structures though it was nearly always covered with some other material. By 1907 the public baths at Hammersmith, London, had been built with rc and by 1908 so had Manchester's YMCA. In 1910 a six-storey office block was erected in Leeds and in 1912 a large office building at London's Baker Street station. In the 1920s and 1930s the most up-to-date rc structures were those designed by Owen Williams who used his engineering skills on many new ideas. But by then new methods had already been worked out in France, Germany and the USA.

France Many French engineers had worked on rc. Even in 1780 François Cointeraux built a concrete house though without reinforcement. In 1832 one was built at Albi by François Lebrun and by 1858 Francois Coignet was given a patent for wrought iron rods in a concrete floor. He knew the iron was in tension. The engineer-contractor François Hennebique who lived from 1843 to 1921 made rc a common material in France. He used steel instead of iron bars and invented the **hook** method of anchoring the steel in the concrete. The hook was made of bars being doubled back on themselves in a hook shape. Hennebique also invented **stirrups** which were loops of thin steel bars wrapped around the other bars to prevent beams from shearing through under their loads.

And it was Hennebique who had the idea of bending the steel bars so that columns and beams and floor slabs could be cast in their wooden moulds as one continuous mass of material. All his ideas are still used today. Auguste Perret took Hennebique's ideas and methods further, and later the International style architect, Le Corbusier, worked for a time in Perret's design office. Le Corbusier himself proposed a rc system for houses in 1914. This was the 'Domino' project where rc floor and roof slabs held apart by columns were the only load-bearing structure. This rc system was to become the basic method of construction for thousands of high-rise housing blocks in the 1950s and 1960s.

USA By the early 1850s Thaddeus Hyatt who lived from 1816 to 1901, was trying out ideas for iron bars in tension zones. Like Hennebique he made **T** beams where a beam and the floor slab it carried were made of the same material, shaped as one continuous unit and he also thought steel better than iron for the reinforcing bars. Hyatt put his ideas into the first scientific book on rc methods.

In 1871 the mechanical engineer, William Ward, tried out ideas for rc and worked on the mathematics for the sag or 'deflection' of beams and their resistance to shearing through when carrying loads. He discovered facts about the stones mixed as **aggregate** into the concrete and about fireproof concrete. At Chester, New York, in 1873 he built a fireproof house of rc which had no wood at all except window-frames, and 15 years passed before anything like it was tried again. Then, in 1902, the rc Ingall's Building went up in Cincinnati.

Up-to-date architects from Europe, such as Walter Gropius, travelled in North America and thought the best buildings were the great grain silos of the Prairies. The silos had strong and simple shapes and were free of ornament. These visitors from Europe also liked the simple shapes, the plain walls and large windows of the rc factories designed by Ernest Ransome.

The first of Ransome's factories went up in 1902 at Greensburg, Pennsylvania. It set the style not only for many other American factories but for the International style itself. The *Beaux Arts* architects both in Europe and the USA liked every building to look like a palace. The International style architects liked every building to look like a factory.

The International style architects used rc because it could give them walls which were thin, flat and smooth, and which could easily be painted white. But because thin rc often cracked badly the architects sometimes used brick infill in an rc frame. When the wall and frame were covered by a coat of cement and painted white nobody could see whether the wall was concrete or not. It looked like rc and that was what the International style architects wanted. (*Figure 17* page 83.)

Many parts of many buildings in the International style were as faked as the *Beaux Arts* palaces. And that faking made later architects look for ways of using the rc structure of a building without covering it by other materials such as the common cement rendering. And a solution to this problem was found, but not until after the Second World War when building began again in the 1950s.

Joseph Emberton's 1931 Royal Corinthian Yacht Club is a perfect example of the International style because it has all the eight main features. (*Figures 17, 17.33 and 17.34.*)

1 **Frame** The lower part of the club building was a rc platform on rc piles on the edge of the River Crouch. The platform could be used like a quay. The steel frame of the building's three storeys rested on the platform and gave the building the International style's airy and open look.

2 **Cantilever** The structural steel uprights were placed back from the outer edge of the floor slabs so that the slabs had a cantilevered overhang. The floor edges seen from outside were white horizontal bands of rc and they were used as viewing balconies. The high gallery of the yacht race Starter's Box was also cantilevered.

3 **Glass** The whole river side of the building had glass walls and glass doors opening onto the viewing balconies. This was 'functional' because people inside could see the yachts and it also showed that the material enclosing the building was only a lightweight membrane and not a supporting wall. The glass wrapped around the corners because they were cantilevered like the corners at the Fagus factory designed by Walter Gropius 20 years before.

4 **Ground Floor** The Yacht Club had the usual void at its lowest level but here the void was also functional because the building was raised about the water level for practical reasons and not just for design effect.

5 **Flat roof** The building was designed in the usual cube shapes of the International style and the sharp edges made against the sky by the flat roof were vital to the cube effect.

6 **Inside-outside** In the bright light of Burnham-on-Crouch the inside and outside of the Yacht Club fused together in the ideal way. People in 1931 seemed

to be less worried than people are today by glare or by heat loss and heat gain through big areas of glass. The style seemed right for building near water such as the Yacht Club or the 1935 De la Warr Pavilion at Bexhill. And this was not by chance, for part of the style's 'new spirit' was to take shapes from modern functional objects like ocean-going liners. The Yacht Club looked like the bridge of a ship and even more so when colourful signal flags fluttered from its great mast.

7 Double-height spaces Changing views through a building were a main feature of many International style designs. Stairs or ramps in mid-air gave people these views as they walked up or down through high spaces in the building. At the Yacht Club Joseph Emberton took this movement outside the building by linking the viewing balconies with spiral stairs which went up to the Starter's Box perched as yet another cube on the flat roof. This gave a sense of being in mid-air between sky and river.

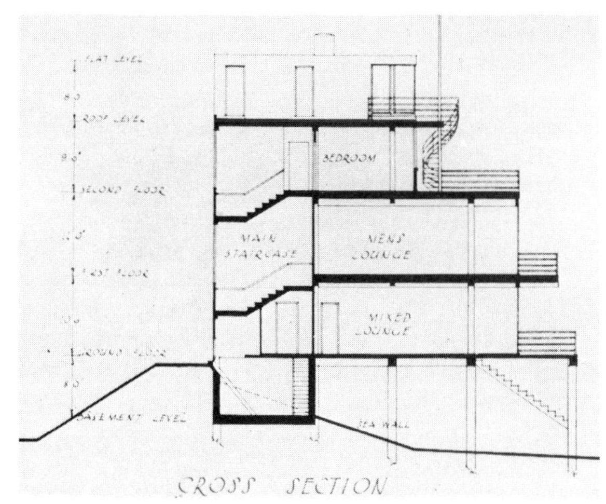

17.33 Royal Corinthian Yacht Club, 1931
Cross-section *Designer* Joseph Emberton

8 Plain and white walls These were another part of the cube effect vital to the style and at Burnham-on-Crouch only the sides away from the river had solid walls. They looked like thin rc but in fact were brick **cavity** walls of the type used for thousands of houses in the 1930s where a 50 mm (2 in.) air gap separated two brick walls linked with metal ties across the gap. The outer wall surfaces at the Yacht Club were covered with white Portland cement. The architect knew that thin rc cracked and so he used brick instead.

Like any other building in the International style the Yacht Club had to look like rc even if this meant faking it. Joseph Emberton was like all other architects in thinking that the International style's looks were more important than its facts.

The Yacht Club looked perfect for its purpose in 1931 and still does today despite the faking. The Club has legal protection from demolition or unsuitable alteration by being 'listed'. It is among the best of Britain's few remaining examples of the International style. A building which was rather like a ship's bridge was just right for a yacht club on a river. But it looked odd in the 1930s and still looks odd today to find two ships' bridges in the middle of London's old Hampstead village and, not far away from the 1938 No 66 Frognal and the 1936 Sun House, was Hampstead's third ship's bridge at Lawn Road where the architect Wells Coates designed a block of flats in 1934. On many sites neither the International style nor its designers seemed able or willing to fit in with the surroundings.

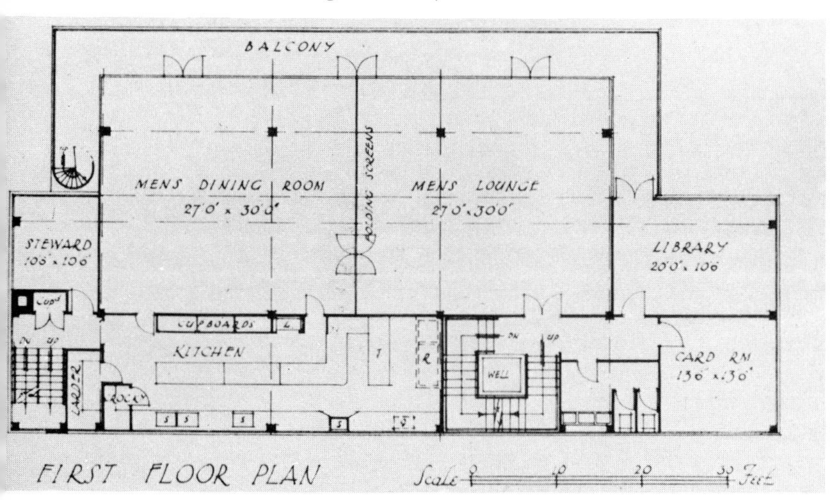

17.34 First floor plan

18 Roehampton Housing

Place Alton East and Alton West, Roehampton, West London
Time *Start* 1952
Finish 1959
Construction time 7 years. In two stages
Purpose Housing
Style 1950s mixed development. High, medium and low-rise blocks

Somers Town and Agar Town were the 1·6 hectares (4 acres) of disease-ridden slums around the open sewer of the Fleet River which were demolished in 1866 to make way for St Pancras Station. They were not London's only slums. Five thousand people died in London during the cholera epidemic of 1832 and another 14,000 in 1848, and London was not the only English city with slums. In Birmingham in 1913 nearly 250,000 people lived in condemned houses of which over 42,000 were without water supply and drains, and 58,000 were without lavatories. (*Figure 18.2.*)

In all the big English cities many people lived in overcrowded, airless, dark, damp and cold houses without basic services and amenities. Lavatories were shared in courtyards, so were the wells and water pumps into which sewage often seeped. Factories belched out sulphur from coal and poured out toxic and foul smelling waste. Houses crowded close to the factories because there was no transport to take the men, women and children to work. At Manchester in the 1840s life expectancy was only 25 years.

The horrors and the marvels of Victorian England existed side by side in London no less than elsewhere and it was in London in the new Houses of Parliament that laws were passed to control the horrors. And it was in the capital city too that the London County Council was set up in 1888 to grapple with the horrors inside its own borders.

18.1 The blocks derived from Le Corbusier's Unité d'habitation at Marseilles

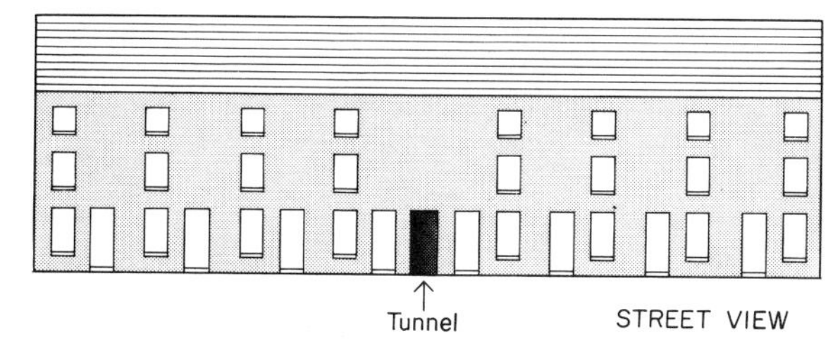

The lavatories are in the courtyard

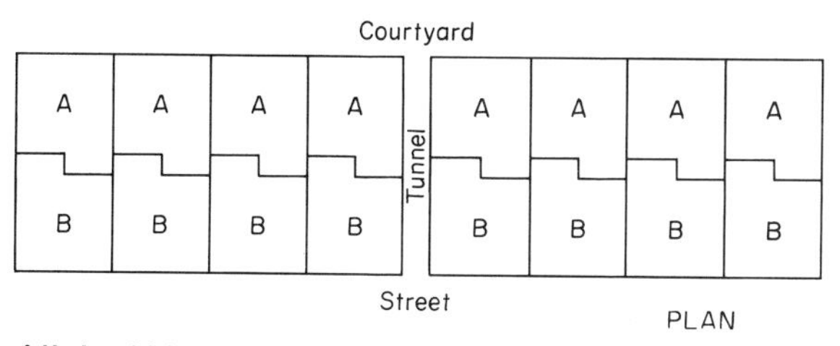

All the 'A' houses back onto the 'B' houses

18.2 View and plan of back-to-back housing before 1875

The new LCC took over the housing schemes which privately financed groups had been building. The aim of such groups was not to make money or even to give the poorest people a minimum standard but to improve their morals. Such groups as the 1844 Society for Improving the Condition of the Labouring Classes asked designers in 1848 to work out schemes of good homes for workers. Queen Victoria's husband, Prince Albert, was the society's president and the idea was that if workers lived in better homes they would lead better lives. The designers' plans showed how six people of the labouring classes could live in 45 sq m (495 sq ft) of space.

In 1841 London had the Metropolitan Association for Improving the Dwellings of the Industrious Classes. The Waterlow Institution was set up in 1863 and by 1874 the Peabody Trust had nearly 4000 people living in its grisly blocks. Harry Darbyshire was the Peabody designer and he planned the first of their blocks in 1864.

In 1851 Henry Roberts had designed ideal cottages for farm workers which were seen by the public when the Great Exhibition was in London's Hyde Park. In 1856 some other homes similar to the cottages were built at Abbot's Langley in Hertfordshire. A basic unit of an open-air staircase giving access to flats on either side of each landing was used for city flats by assembling the basic units side by side in long terraces and by piling them one on top of the other five storeys high.

There was a Housing Act in 1890 which helped the LCC when it took over London's housing problem from these private groups and trusts. The key to housing was land. In the 1890s the LCC owned two kinds of land: built-up areas and green areas.

1 Built-up areas

These were sites in London's centre such as the vast East End Dock areas along both banks of the River Thames. But land owned by the LCC did not amount to much and central area sites were expensive to buy. So the LCC tried to get a large number of people into its housing blocks.

In 1895 the Boundary Street Estate was opened at Shoreditch. In the slums that had to be cleared before building could begin it had been usual for one child out of every four to die young. And Boundary Street was only one of many such desperate areas cleared by the LCC. The people had to be re-housed on the same site, so well over 1000 new homes were built in blocks five storeys high without lifts.

To make the buildings look more friendly and human than the grim blocks put up by the private trusts, the LCC architects used ideas from the Arts and Crafts style which were first seen in the 1880s.

Between 1899 and 1903 the LCC built the Millbank Estate at Westminster. The old Millbank prison was knocked down and 17 Arts and Crafts blocks of five storeys were built on the 3·6 hectare (9 acre) site.

By 1930 LCC designers had worked out a standard five-storey block for built-up areas. The ideas for these blocks of small flats were taken from large houses designed in the Classical style of the 1700s. The LCC was building with this fake Georgian style at the same time as Joseph Emberton was using the International style for the Yacht Club at Burnham-on-Crouch. The LCC's long Georgian blocks were built around bare and noisy courtyards overlooked by access balconies. In 1929 a big estate of that sort was built at China Walk in Kensington.

By the start of the Second World War in 1939 nearly 89,000 families lived in new LCC homes. And by the end of the war in 1945 some 89,000 LCC homes were damaged, 2500 of them beyond repair. So the LCC set a target of 100,000 new homes to be built quickly, and by 1950 32,500 of them had been built. The designers used ideas from Sweden so many of the buildings went up to nine storeys and had lifts. They were built with standard reinforced concrete components made off-site in factories. So with these up-to-date methods of industrial production the buildings went up rapidly. Some of them were in London's East End at the Woodbury Down Estate which people moved into in 1948.

Speed and low cost took over. The buildings and the estates became uglier and the flats became more inconvenient and uncomfortable until by the 1950s a new start was made when the LCC Architects' Depart-

18.3 Towers at Alton Estate West during construction in 1957

ment took over from the Housing and Valuation Department.

The idea of 'mixed development' was then taken up which was to put two-storey houses and medium-height blocks of maisonettes and high-rise blocks of flats on one site in hopes of mixing the types of people living there and of making the surroundings varied and interesting. The LCC's first mixed development was the Alton East Estate begun at Roehampton in 1952. (*Figures 18.1 and 18.2.*)

Planners, building committees and architects thought that in mixed development they had found the perfect way to build homes and after Roehampton the LCC put up many mixed schemes. In some of these the tall blocks went up to 25 storeys. In 1965 the LCC became the Greater London Council (GLC). They had been building 4500 homes every year, only 500 short of their target.

The Canada Estate in the built-up area of Southwark opened in 1963 and though there were already warning signs that the tenants did not like living in such high blocks nobody in the GLC seemed to take any notice and they continued to build them. The government made local authorities, such as the GLC, look at building methods which used factory-made components and mechanised methods of building on site because these 'industrialised' methods were quick. So the GLC put its target up to 7750 new homes a year. Half were to be built by 1970 using industrialised methods.

At Westminster's 1968 Elgin Estate the 21-storey blocks were finished outside with glass fibre panels which was an industrial material given a new use as the cladding of a structural frame. But in spite of industrial methods and materials the end of high-rise building was in sight. The technical faults of badly built blocks were already showing how costly the future maintenance of high-rise blocks would be; and vandalism had begun. People were cut off from social contact, and the open spaces between the blocks were unsafe and derelict.

In 1967 the government took more control over the financing of local authority homes. This led to buildings with fewer storeys. In 1967 the standards for bigger and better rooms set out a few years earlier in the official 1961 Parker Morris Report became compulsory for local authority homes. But those standards could not be applied to high-rise blocks.

By 1970 the GLC had tried other ideas. By putting lower buildings close together the architects could still keep high numbers of people on an estate. This made the tenants' surroundings better than in tall blocks both indoors and outdoors. Another new idea was to bring old houses up-to-date. The GLC's first big up-dating scheme was done in 1971 at Porchester Square in Westminster. By 1980 this method of re-using old housing stock had been taken up by local authorities and private firms all over Britain. By 1982 the government had made local authorities build new homes for sale rather than for letting and this changed the whole basis of local authority housing. For a hundred years it had served people's needs, but now it was to suit their pockets.

Some buildings that went up quickly in the 1960s and 1970s came down even more quickly in the 1980s when some London authorities used explosives to demolish high-rise blocks that were not yet 20 years old. Many local authorities all over Britain found themselves with mounting maintenance costs, first for decaying old property and secondly for new housing that was falling to pieces.

By the 1980s there was a shortage of land for housing.

But the built-up areas of large cities still offered local authorities infill sites and London had large sites in Dockland because the docks had been re-sited down the Thames Estuary where container ships could berth.

2 Green areas

From 1888 when the newly set up LCC was looking for housing sites to buy they found the green areas of London's fringes cheaper than built-up areas. On green sites the homes could be houses with gardens rather than flats in big blocks, and they could be grouped like country cottages with grass and trees. In 1912 Old Oak Estate was among the LCC's first of this sort, and in 1921 people moved into another called the Roehampton Cottage Estate. By then for the first time in Britain the government had given grants to local authorities for housing.

The LCC went on building in its green areas. In 1956 they put five-storey blocks and small, two-storey houses at Blackheath's Brooklands Park Estate making a countryside effect which was used later in many other green areas. And as large building sites became scarce and costly in the 1970s the new GLC began its infill methods by which small new houses were slotted into gaps between older buildings. This also helped to retain the features and character which people liked about an area. One example was built in 1976 at Ruthin Road in Greenwich. By the 1980s the GLC architects were designing clusters of one, two or three storey houses on sites not thought suitable before because of their odd-shaped boundaries.

The finest of all the green areas in London was the land at the edge of Richmond Park where the East and West Alton estates were built between 1952 and 1959. The LCC had 52 hectares (130 acres) of parkland on which to put over 2600 homes. These could not be like the Roehampton Cottage Estate built 30 years earlier because 2600 houses would have been even closer to each other than suburban semi-detached houses. So the LCC architects put two ideas together. They designed a mixed development of the high, medium and low-rise sort used in built-up areas but placed it in the Roehampton green area.

The idea that 'green-is-good' was not new. For 150 years there had been attempts to build ideal towns in green surroundings. In 1800 Robert Owen built New Lanark in Scotland as a new settlement for his cotton mill and its workers. And among other companies which made new towns for their workers was the Butterley iron company that built the 1866 roof of St Pancras Station. In 1850 the company re-housed its workers in the specially built Ironville near Derby. And in 1851 the woollen manufacturer Titus Salt took his mills out of Bradford in Yorkshire and put them and his workers in a new town called Saltaire.

The Halifax Building Society grew from people's ownership of their homes built at Akroyden in 1861. In 1879 George Cadbury's cocoa and chocolate firm began the new town of Bournville and this was followed in 1888 by Port Sunlight in Cheshire built by W H Lever for his soap factory and workers.

By 1898 Ebenezer Howard had set out his ideas for clusters of **Garden Cities**. These were to have the best features of both town and country but none of the worst features. People would live in pleasant green areas which would be smoke-free because industry was sited away from houses. Every one of the garden cities would have its own greenbelt and all would be linked by trains and canals. When the numbers of people reached 30,000 another new garden city would be built nearby.

In 1903 Howard set up a company called First Garden City Limited which bought over 1800 hectares (4500 acres) at Letchworth in Hertfordshire where the architects Raymond Unwin and Barry Parker designed and built the first garden city for 30,000 people on 600 hectares (1500 acres) using the remaining 1200 hectares (3000 acres) for green areas such as farms and green belts. A second garden city was begun at Welwyn in 1920.

These two firmly planted the green-is-good idea in planners' and architects' and private developers' minds. It became official government policy after the 1918 Tudor Walters Report which said that the big cities should re-house their people in estates of two-storey houses with front and back gardens and indoor bathrooms and lavatories and with only 30 houses per hectare (12 per acre). This was the start of the vast suburbs of 'council houses' and of privately built semi-detached houses which sprawled out from cities in the 1920s and 1930s. And the New Towns built after the Second World War were also based on the idea that green-is-good despite the fact that for many former city-dwellers green was also boring.

The designers of the two Alton estates at Roehampton tapped another source for their idea of mixed development in a green area. And this was the work of the French architect Le Corbusier. In a 1922 exhibition he showed his ideas for a 'Contemporary City'. Its three million people were to live in high-rise blocks set amid greenery through which fast motorways led to the city centre with its 60-storey towers. Le Corbusier made a city plan for Moscow in 1933 which had the same idea of high-rise blocks in green areas. In 1952 Le Corbusier finished a huge high-rise block at Marseilles in the south of France.

This French block re-housed the shipyard workers from the Old Port area of the city damaged in the Second World War. The vast 'Living Unit' (*Unité d'Habitation*) housed 1600 people in its 20 storeys and gave them shops and community facilities all under the one flat roof which was used as an outdoor play area. The immense building was 140 m (460 ft) long and an 'inside street' (*rue intérieure*) at every third floor level gave access up or down to large flats. Some of these flats went right across the Living Unit's 24 m (79 ft) width and had double-height living-rooms looking over the countryside.

The ground level void of Le Corbusier's 1920s and 1930s buildings in the International style was used again on a massive scale at his Marseilles Living Unit so that the landscape was not blocked off. But this Living Unit was not in the International style because it was designed on grids of Golden Rectangles as was his 1927 house at Garches near Paris. Le Corbusier had invented his own system of golden ratios and he called it the **Modulor**. All the prefabricated parts of the Living Unit were designed with shapes and sizes taken from the measuring scales of the Modulor. The parts were made with reinforced concrete. But it was not the smooth and white rc or fake rc of the International style. It was concrete made to look heavy and left with the rough surface of the wooden moulds the concrete was poured in.

This heavy and rough **brutal concrete** (*beton brut*) was soon used as a style by architects all over the world. Rough concrete was a main part of the 'New Brutalist' style. But many people who at the time thought the Living Unit at Marseilles one of the greatest marvels of modern design were later to think of it as one of the greatest horrors.

The Living Unit had all the features that designers of later high-rise blocks were to copy and often copy badly. The empty ground around the block, the empty ground floor, the bare concrete and buried corridors, the danger of unguarded stairs and lifts, the loss of mother's control over children playing 20 storeys down at the ground and a layout which stopped any growth of community feeling that streets allow, were all features that by the 1970s brought high-rise buildings to an end. Tall blocks were a social disaster that had wasted thousands of millions of pounds of public money and is still costing millions in demolition or endless maintenance.

In 1952 Roehampton became a mini Marseilles. In fact the London green spaces got no less than five Le Corbusier-type Living Units though they were only 11-storeys high and much smaller in every way. Besides these there were 11-storey towers, four-storey blocks of maisonettes and three-, two- and one-storey houses with gardens all spaced out in the parkland and its trees. Half of the new homes went into the 30 Alton high blocks.

All the high buildings had rc frames and as the frames went up the floors and stairs and outside wall panels were hoisted up and fixed in position. They were ready-made as **precast** components. Like the cast iron components used in buildings a hundred years before, rc components were cast into specially made moulds and when the concrete had hardened they were taken out of the moulds and fixed in the building. This method of precasting concrete could be done by a specialist manufacturer in a precasting factory so that each precast item such as a staircase or a floor slab could be produced under conditions of industrial quality control.

In large projects with large numbers of repeats the method was economical because the moulds were used many times. To save transport costs a precasting plant was sometimes set up on the building site itself. When the mould was built up in the position and shape of the building structure itself the concrete frame was called *in situ* rc. It was not unusual for an *in situ* rc frame to have floors and roof, staircases and outside panels all of precast units. The size of precast components was limited by transport and site-handling capacity.

Speed and economy of construction on site were greatest when the buildings were in long straight rows so that the cranes had a maximum of reach with a minimum of relocation. By 1970 the 2000 homes of the GLC's Aylesbury Estate had gone up at Southwark with endlessly repeated precast units. This sort of large-scale building operation in built-up areas was called **comprehensive redevelopment**. The speed with which the endlessly repeated concrete units went up was only matched by the speed with which the endlessly repeated vandalism began.

19 West Bridgford Boys School

Place West Bridgford, Nottingham
Time *Start* March 1960
Finish September 1961
Construction time 18 months
Purpose A technical school
Style System-building. Sometimes called 'prefabricated' or 'industrialised building'

The ready-made structure is one of the oldest ideas in building. When the Normans landed in England in AD 1066 they brought with them a prefabricated fort. The wooden parts came in their invasion ships already cut to shape and with holes drilled ready for the pegs that fixed the pieces together. The fort was started on the morning of the invasion and it was ready by nightfall.

Wooden building kits in Europe were prefabricated from at least AD 1200. During the 1390s the hammer-beam roof for Westminster Hall was made in Surrey

19.1 Courtyard view under the library to the gym. The steel stanchions, tile-hanging and weatherboarding are all part of the Clasp system

and taken by cart and boat to London. In the 1500s ships carried ready-cut and numbered stones as ballast when sailing from Portugal to Brazil where the stones were unloaded and assembled on site into church buildings. In 1624 a ready-made wooden house kit was shipped from England to America.

By 1790 lock-keepers' houses by canals were ready-made, built of iron plates and by the 1820s kits for whole buildings were being prefabricated for people going to tropical countries from Britain. In the late 1820s a Mr Manning of Holborn in London sent a prefabricated wooden house to the island of St Helena. Mr Richard from the East End was the carpenter of the 36 m × 15 m × 7 m (118 ft × 49 ft × 23 ft) house which was for the captive Napoleon to live in. Mr Manning made a thriving business out of supplying prefabricated houses of all sizes.

In 1829 Richard Walker made the first practical wrought iron sheeting and in an advertisement pointed out how good this corrugated iron was for exporting from Britain. It was easily packed and stacked and took up little storage space. It was easy both to put up and to take down or move from one site to another.

By the 1830s many firms were exporting ready-made iron building kits. In 1843 William Laycock of London sent an iron palace to the Calabar River in Africa for King Eyambo and his 320 wives. (*Figure 19.2.*) The firm of Wood and Weygood were shipping kits for warehouses and dwellings to the West Indies as well as Africa. Iron churches went off to Britain's colonies. One that travelled furthest was the iron front for the church in Sydney, Australia. In 1846 Barbados got an iron sugar factory and in 1848 an iron market building was shipped to Trinidad.

Prince Albert ordered a corrugated iron ballroom for Balmoral Castle after he saw a building like it which E T Bellhouse had put on show at the 1851 Great Exhibition. And the best of all prefabricated buildings was the Crystal Palace which was not only put up in Hyde Park but taken down after the exhibition and re-built at Sydenham in south-east London where it stood until destroyed by fire in 1936.

19.2 King Eyambo's Palace

Today people would say all those wooden and iron structures had been **system-built** and were **system buildings**. This simply means that the designing, making and assembling the buildings and their components on site is considered as a complete sequence of operations right from the start. The whole process is a **system**. A Red Indian wigwam was a system building and so was the West Bridgford Boys' School begun at Nottingham in 1960. A vital difference between a wigwam and the school was that, while the wigwam was hand-made from natural materials, the school was made by machines in factories from machine-made materials. Only a minimum of manual work went into the assembly of the factory-made kit on site. (*Figures 19.1, 19.3 and 19.4.*)

Ever since an 1876 Act of Parliament made it compulsory for children to attend school between the ages of five and ten the design of school buildings has been based on standard types approved by government authorities. (*Figure 19.5.*) After 1876 the local School Boards put up brick buildings with classrooms grouped around a central hall on each of two or three storeys and many of these Victorian schools are still in use. In 1902 the Local Education Authorities took over from the Boards and a medical service for schools was set up in an attempt to deal with the tuberculosis, diphtheria, rickets and other diseases common among children in those days. More light and more air made school buildings healthier after 1902.

Between 1907 and 1926 there were special building regulations for schools which allowed cheaper construction. The 1920s saw schools with wood and glass doors completely opening-up the opposite sides of the classrooms which also had high level windows. These **open-air** schools were one-storey structures which needed a good deal of land, so by the 1930s many new school buildings had a two-storey layout of rooms grouped around large open-air courtyards or **quadrangles**.

Some quadrangle schools were in a rigid and fake Classical style and some in an equally rigid and fake 1930s modern style copied badly from the brick buildings of the Dutch architect Willem Dudok. His 1926 Vondel School at Hilversum had ideas which English designers were using ten years later. The rigidity of these English school buildings was not just a matter of design but also of the rigid education programme.

That rigidity of design and of teaching was swept away by the 1944 Education Act which gave all children from all levels of British society equal access to an

19.3 Looking from a dining-classroom across a courtyard to the double-height assembly hall

education that matched their talents and their aptitudes. The 1944 Act helped new and creative ideas about teaching and needed new types of buildings. The new type of school needed rooms which were not only open to light and air but could also open into one another with access to grass and trees. The garden city had given people a new way of living and now the garden school gave children a new way of learning.

The schools created by the 1944 Act were designed to let smaller groups of children and staff make their own surroundings in their own part of the building. The furniture and equipment was designed for multiple use. The Victorian idea of 'classrooms' did not exist any more and instead there were spaces which could be used for many different purposes in one school day. And as children grew older they used rooms designed for the teaching of music or science or sports.

Architects saw that they had to design school buildings which let teachers and children get together freely in groups. And these groups had to be able to use their tables and chairs and their pin-up wall surfaces in any way that related to their learning activities. Lightweight outside walls with large windows and doors opening onto paved areas outside seemed to solve the design problem. And a lightweight framed structure seemed to solve the construction problem. A large number of new schools were needed quickly as a result of the 1944 Education Act. System building was the answer.

Hertfordshire was the first county to apply the methods of system building to schools and in 1947 the county's architect Charles Aslin and his team built a new school at Cheshunt. It was the first prefabricated school in Britain. And in 1956 Nottinghamshire County Council set up a special group for designing and building schools and later were joined by Derbyshire and the City of Coventry councils.

These three local authorities pooled their resources of technical and management skills in a group or 'consortium' and worked out a special programme of system building for all the new schools needed in their three districts. They called it the 'Consortium of Local Authorities Special Programme' or CLASP for short. The system the designers invented itself became known as the **Clasp system**. By 1962 when the West Bridgford school was built the Clasp programme was dealing with projects worth nine million pounds. This was big enough to allow four members of the design

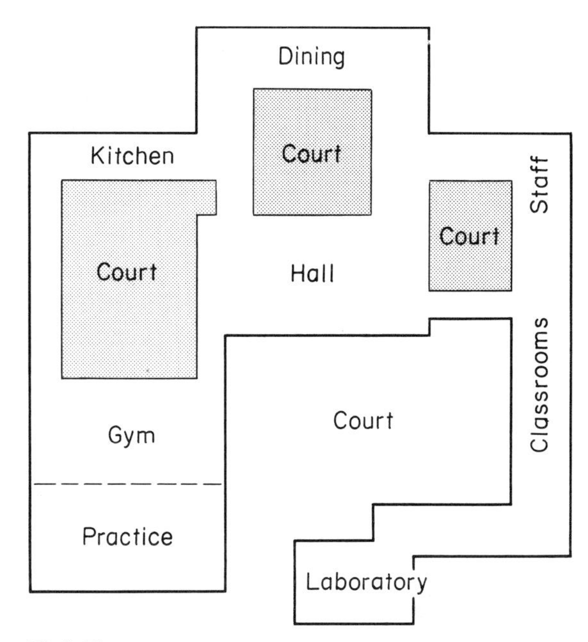

19.4 Plan
19.5 Plan of earlier schools from 1880 to 1960

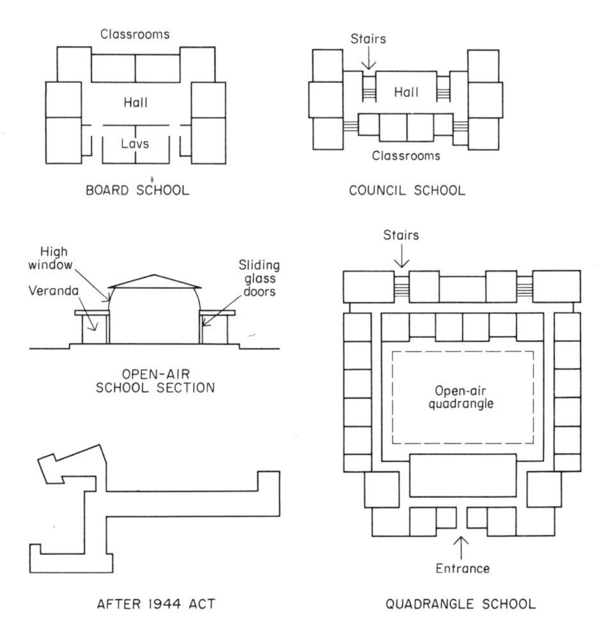

team to work full time on research and development of the system.

Many sites in the Midlands area around Nottingham, Derby and Coventry were liable to subside because of the coal mining. So the Clasp system was designed to allow settlement movement in the buildings. The system had a lightweight steel frame with hinged joints and diagonal wind bracing joining opposite corners of the frame to give stiffness to the structure. But the bracing had a metal spring so that the hinged frame could deform slightly.

The light steel frame supported steel lattice beams for the roof and upper floors. Prefabricated wood floor panels rested on the lattice beams. The frame also supported the prefabricated panels of the outside walls. There were precast concrete panels with their small stones exposed on the surface. And there were panels of traditional **weatherboarding** which were wooden boards lapped and fixed together horizontally. And there was **tile hanging**, a method of hanging clay tiles vertically on an outside wall which craftsmen had done for hundreds of years on country cottages and farm buildings. The windows and doors came ready-made from the factory and complete with their glass fixed by gaskets like those used for car windows. Pipes and connections for the plumbing also arrived at the site already jointed and ready for fixing into the building.

These Clasp components together with all the inside floor, wall and ceiling materials could be built up to four storeys high and Clasp was soon seen by everybody to be such a good system that it also began to be used for other types of local authority buildings.

System building differed from craft building in one vital way. When all the parts of the Clasp kit arrived on site from many different factories each part was in a finished state. So each part had to be exactly the specified shape and size and had to fit exactly with the other components next to it. By craft building methods a carpenter could plane down a piece of wood that was too big or a plasterer could cover brickwork that was uneven. But with system building methods no adjustments could be made on site. The process was one of industrialised assembly.

This one factor of assembly determined the whole process of design, component manufacture and site assembly. A grid of squares was the key to the process. Each component was given its own numbered space in the grid. It had to fit inside that space and not overlap the spaces next to it. Tolerances were allowed for slight variations in the size of the components which always occurred in the manufacturing process and other tolerances were allowed for slight variations in the setting-out of the building on site. So the component was slightly smaller than the space given to it.

The spaces were in multiples of a basic **module** measuring 100 mm (4 in.). **Modular co-ordination** was the use of the basic module to relate all the grid spaces and components. So this enabled designers and manufacturers and site assembly teams all to work together within the co-ordinated system. And once the Clasp system had been set up buildings were designed and erected quickly because the same kind of kit was used for every project.

Work began on site for the West Bridgford school in March 1960. A 127 mm (5 in.) thick concrete slab reinforced by a mesh of small diameter steel rods was laid for the school's ground floor of 2590 sq m (27,913 sq ft). This rc slab was a **raft** designed to resist cracking if coal mines under the site made the ground move.

The prefabricated parts of the Clasp kit were ready for assembly. For its single-storey areas the school had steel stanchions with a hollow section 114 mm (4·5 in.) square. These stanchions were **cold formed**, a process by which a machine bent strips of sheet steel into the required shape. Each of the stanchions took up its place in the grid which had its controlling lines at every 10 modules – 1·016 m (3 ft 4 in.). Some stanchions were spaced out on every other grid line so they were 2·032 m (6 ft 8 in.) apart and others were on every third line at 3·048 m (10 ft). These stanchions were welded to the steel lattice beams of the roof. In single storey Clasp buildings the beams could span up to 160 modules, 16·25 m (53 ft 4 in.).

For the school's two-storey areas, stanchions of the same size were used but some of them were **hot-rolled** sections which came from the steelworks ready-made as tube sections. Clasp's longest span for beams in two-, three- and four-storey buildings was 90 modules, 9·144 m (30 ft).

The stanchions were fixed by steel plates to the rc raft slab and a 100 × 50 mm (4 × 2 in.) steel tie ran along horizontally from one stanchion to the next giving them stability. The tie was a **channel** section 6 mm (0·25 in.) thick. The stanchions also carried steel **L** rails from which the outside wall panels were hung.

Clasp was called a **closed system** because it only used components specially designed and manufactured for it. But there were **open systems** which could combine their own special components with other building materials and products. Twenty years after the West Bridgford school opened the whole building industry was using many more factory-made products. Buildings had come to rely entirely on the products of industry. By the 1980s construction and services for certain types of building had merged with the high technology of industrial processes themselves. One of Britain's first **high tech** buildings was the Inmos Micro-chip factory. The Age of the Micro-chip had a design style all of its own.

20 Inmos Microchip Factory

Place Newport, Wales
Time *Start* January 1981
Finish April 1982
Construction time 15 months
Purpose A factory for micro-electronic circuitry
Style High Tech

The **high technology** which gives the style its name was the combination of industrialised building materials and methods of erection with a complex network of services. The High Tech style exposed its basic structure which was often a steel framework and also the machinery and pipes of the services such as air conditioning and ducts. The frame and ducts were painted with bright colours and made a spectacular display of up-to-date construction technology. The inside spaces of High Tech buildings usually had a high level of control over the air and light. For example the Inmos factory had to have spaces which were as free of dust particles as any man-made spaces could be. (*Figure 20.1*)

20.1 Masts and tubes. The High Tech display of structural steel and service ducts

Buildings had always controlled their air and light either by design or by chance. Sometimes the control was good for the people inside and often it was bad. The desert tent and the Arctic igloo, the African straw house and the Japanese paper house were all a response to local climate made by readily available methods and materials.

Comfort in buildings is partly to do with people's bodies and partly with their minds. In the 1880s people accepted conditions which people in the 1980s would not. The damp and draughty houses the Victorians thought normal were to be replaced by today's double-glazing and central heating. Many Victorian medical men knew that lack of good ventilation helped the spread of disease and not only in the slums but in larger houses too.

Dr John Hayward knew this in 1867 when he built a house called The Octagon in Liverpool's Grove Street. (*Figure 20.2.*) He designed this house for himself with a built-in system of air change. To make air flow through his house the doctor simply used the fact that hot air rising up a flue drew cold air in at the bottom of the flue. This was the basic idea of the ventilation system in the recently completed new Houses of Parliament.

Dr Hayward wrapped a second flue around his kitchen flue so that there was one chimney inside the other with an air gap between them. Smoke from the kitchen fire went up the inner flue and hot air went up the outer flue. This outer flue drew cooler air through specially built slots and ducts everywhere in the house and the resultant air flow caused air changes in all the rooms.

The air movement drew air from outside the house into a basement room where much of the soot in it could fall to the ground. The air was then drawn over 25 mm (1 in.) diameter hot water pipes and rose by convection through floor grilles to a central lobby on each of The Octagon's four storeys. The soot-free and warmed air then flowed from the lobbies into the room through rows of 175 mm (7 in.) high openings.

The air was heated to make convection currents for air change and not so much to warm Dr Hayward's family who still had to sit by coal fires like most other Victorians. But they could watch the naked flames of the gas lamps flicker as the warmed air was drawn out of the room through a zinc duct over the lamp taking some of the gas soot with it. The air went on up through ducts built in the walls, being sucked all the time by the heat siphon above the kitchen fire. At last the air reached the kitchen flue and was exhausted into the open air. The whole of this ventilation was done by convection. Hot air flowed upwards drawing in cooler air. The system had no mechanical aids such as pumps or fans.

Air changing also concerned the designers of large public buildings. Their main method was like Dr

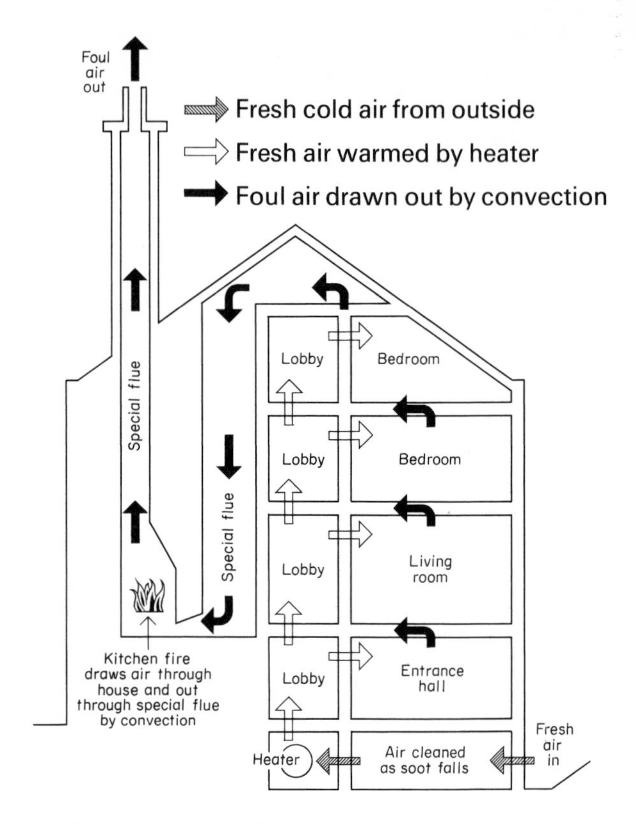

20.2 Dr Hayward's 1867 house, The Octagon

Hayward's house and they made huge volumes of hot air flow up specially built brick flues. A siphon extract system of that sort was used in Manchester's Free Trade Hall and for the 1885 Leeds Town Hall where the architect, Cuthbert Brodrick, made the big siphon ducts into outside features of the building. Such volumes of moving air could not be properly cleaned of pollution nor was there any kind of control over the amount of moisture in the air. Such air-change systems were far from good but they were better than nothing for crowded audiences at public meetings and concerts.

A better way to move air was by a large fan turning on an axle. In 1736 the old House of Commons used to get some sort of air change from a fan made to go round by a man turning a handle like the system which forced air into the lower decks of sailing ships at that time. But it was not until after 1860 that fan-forced air changes become common in buildings. At first the fans were turned by steam engines and later by slow-turning gas engines run off town gas mains. Such fans were fitted at first in coalmines and steamships and for industrial processes such as drying tea.

By 1870 the Sturtevant Company in the USA got a patent for their system in which a fan blew air across a coil of hot pipes and by 1900 high speed fans were available. Architects soon installed them in non-industrial buildings such as the opera houses in Vienna and New York which both had fan-forced ventilation systems. The fans of a **plenum** system filled a whole building with air under pressure which then forced its way out through openings.

The 1906 catalogue of the American Sturtevant Company had a picture of a fan blowing warmed air into a big shop in Boston and another picture showed the fan and ducting in the basement of a school in Michigan. By 1910 fan-forced air changes had become a main feature for large buildings of all types along with the warming or cooling of the air.

For thousands of years people in cold climates had no means of heat other than the open fire. And in hot humid climates people tried to keep cool in open-sided houses which let air through while a thick thatch kept off direct sunshine. In hot dry climates people kept cool behind thick walls and roofs of sun-baked mud or mud bricks. High rooms and small high-level windows caused up-draughts which helped to evaporate moisture from people's skin and so made them feel cooler. In India's hottest places there had been a simple fan system for hundreds of years. It was a mat hung from the ceiling and swished slowly to and fro by hand.

All these old methods of heating or cooling needed buildings of the right shape and materials. In Britain the main heat source changed from wood to coal. Londoners used so much coal that a tax on it paid for a substantial part of St Paul's Cathedral when it was being built between 1675 and 1710, though the building itself had no heating system for a very long time. And by 1866 when St Pancras Station was begun Londoners were burning four million tonnes of coal a year much of which came into St Pancras from the Midlands coalfields. Coal fires gave out heat from only one place in the room and were no good for large buildings.

A **heating system** is the transfer of heat from its source to an outlet other than the source. The Romans let hot gases and hot air flow through the walls and under the floors of their buildings by hypocaust systems such as those at Fishbourne in Sussex. The hot water for their baths flowed short distances through pipes from bronze boilers. In the 1400s there were ideas about carrying hot water through pipes to heat rooms but no installations resulted from the ideas.

In 1742 Benjamin Franklin invented a stove for heating American houses by convected air rising from a metal box around the stove. The steam engine's inventor James Watt heated his office in 1784 by steam and in 1804 the Houldsworth textile mill in Glasgow was heated by steam passing through hollow cast iron columns.

By 1860 hot water heating systems were in general use for large buildings. Designers and pipe fitters could make quite good installations. At first the systems worked by convection. Hot water went up, was cooled and then came down again so causing the water to circulate. But radiators on the cooler, return part of the circuit were often cold so later systems had pumps to push the hot water around the building.

Fans which heated buildings by forcing warm air through ducts could also cool buildings by forcing cold air through the ducts. In 1880 the audiences in New York's Madison Square Theatre were kept cool by the warm outside air being forced by fans to flow over blocks of ice. In summer four tonnes of ice would be used in just one evening. Some buildings were cooled by air passing over pipes chilled by refrigeration.

The mechanical equipment for air control took up a great deal of space. Fans had to draw fresh air into the building, blow it through a heating or cooling chamber and then force the air through ducts to every room in the building. Then the stale air had to be drawn out through extract ducts and sometimes this was done by a different group of fans. The fans and their motors and the heating and refrigeration machinery were all heavy and so were put in basements though some fans went onto flat roofs. The ducts had to run vertically and horizontally through the building. Duct cross sections were large near the fans but got smaller and smaller towards the end of the duct run as smaller ducts branched off.

Architects and structural engineers had to provide space for all this ducting. In industrial buildings the plant and its ducts were planned as part of the factory machinery. For example fans and ducts for the extraction of sawdust from the woodworking machines of a furniture factory had to be designed in relation both to the building and to the machines in the building.

In non-industrial buildings people did not want to see the plant and its ducts and pipes so designers had to hide them in special spaces designed for the purpose. The location of the plant and ducts had to ensure efficiency for the proper working of the system and had to be as economical as possible. The systems were designed by mechanical engineers who specialised in heating and ventilating installations and they co-ordinated their schemes with the architect's design for the building as a whole.

Designers soon began to see that the new air control systems could have an effect on the shape of buildings. Good air could now be taken to any part of the building. Windows were no longer needed to provide air changes. Any space in a building could be ventilated, heated or cooled without windows playing any part in the system. This fact was to produce entirely new shapes for buildings.

The American architect Frank Lloyd Wright was one of the first to see the new possibilities of design arising from mechanical control of air. His 1906 Larkin

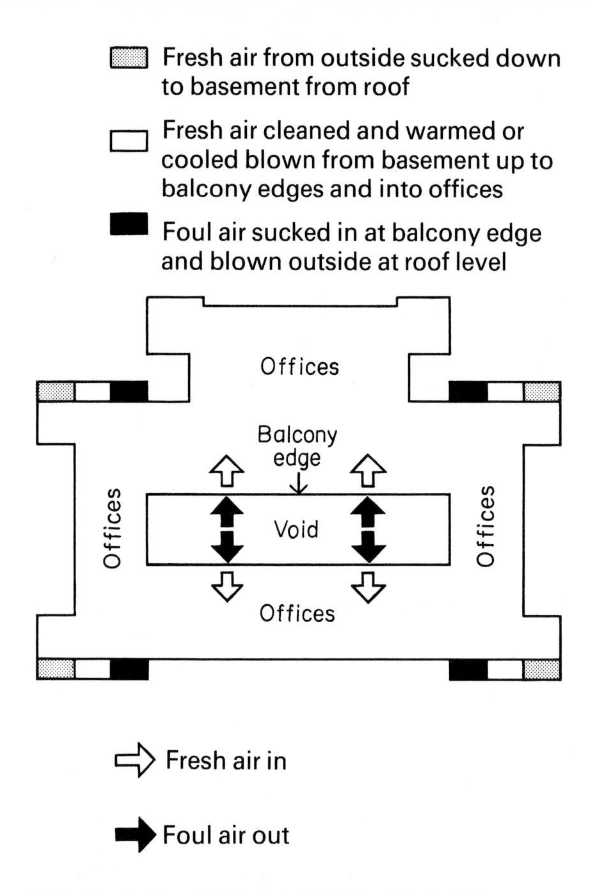

20.3 Frank Lloyd Wright's 1906 Larkin Building

office building at Buffalo had four storeys of offices built as wide balconies overlooking a central space going right through the building from top to bottom. (*Figure 20.3.*)

Main air ducts were put in special towers at the building's corners. Fresh air was drawn in at the top of the towers well above the level of dirty street air below and then was drawn into the basement for cleaning and warming. After 1909 air was also cooled there in summer. Then fans forced the treated air up other ducts to outlet grilles all along the front edges of the balcony offices where stale air was extracted. The windows were closed so that they let light in but not dirt and noise.

The Larkin Building showed what new kinds of layout and space could come from using mechanical air control in a simple and practical way. The system had to be part of the design right from the first ideas for the project. The Larkin Building's strong block shapes and horizontal rows of windows had no need of Classical Orders or Gothic Revival decoration. Frank Lloyd Wright thought those styles belonged to a past that had no place in the America of 1906.

Yet in spite of these mechanical aids the control of air was still not total. Control over the air's moisture had still to be invented. The Larkin Building was as good inside as any building could be in 1906 but it was not fully air conditioned. For air conditioning the system has to have a way of putting moisture into dry air and of taking it out of damp air.

It was also in 1906 that the American engineer Willis Havilland Carrier applied for patents. He became one of the main inventors of air conditioning and he lived long enough to see his 'Conduit Weathermaster' system installed at the 1950 United Nations building in New York. In the USA's extremes of climate, air conditioning was to become part of the American way of life. The technology of miniaturisation brought air conditioning to people's homes as a piece of ordinary domestic equipment after the Second World War.

One of the first air-conditioned buildings anywhere in the world was the 1903 Royal Victoria Hospital in Belfast where the ventilation methods used in ocean-going liners were applied to the building. Belfast was a famous shipbuilding city and its new hospital was one of the first to have its layout fixed by the air-conditioning system. The wards and operating theatres and other medical rooms were placed side by side in a long row. There were no windows in the walls except at the end of each ward. Instead air came in through high-level ducts and went out at floor level and so to the outside air. (*Figure 20.4.*)

The hospital's 17 wards were laid out like the cabins in a ship. At one end of the building two 3 m (10 ft) diameter fans sucked air from the outside and forced it over heating pipes and then into a 6 m (20 ft) deep duct that ran under the floor along the whole length of the 152·5 m (500 ft) long ward block. From this main duct smaller ducts branched off to supply each of the 17 wards. At its far end the main duct was reduced to only 1·8 m (6 ft) deep.

Moisture was controlled by ships' ropes hanging behind the intake grilles. The ropes made a filter through which the outside air had to pass. The ropes were kept wet by a spray so the air left its soot behind and picked up moisture on the way through. A maintenance engineer controlled the amount of air moisture by varying the spray on the ropes. He kept daily records and they later showed how carefully the whole system had been run.

Graumann's Theatre of 1922 in Los Angeles was the first to be fully air conditioned by mechanical equipment and the 1928 Milam Building in San Antonio, Texas, was the first office block with air conditioning. But it was 1950 before the design problems for air-conditioned buildings were solved.

One of the factors which affects the amount of moisture in the air is the air's temperature. Warm air holds more moisture than cold air. Because of the heat from the many light bulbs in office buildings of the

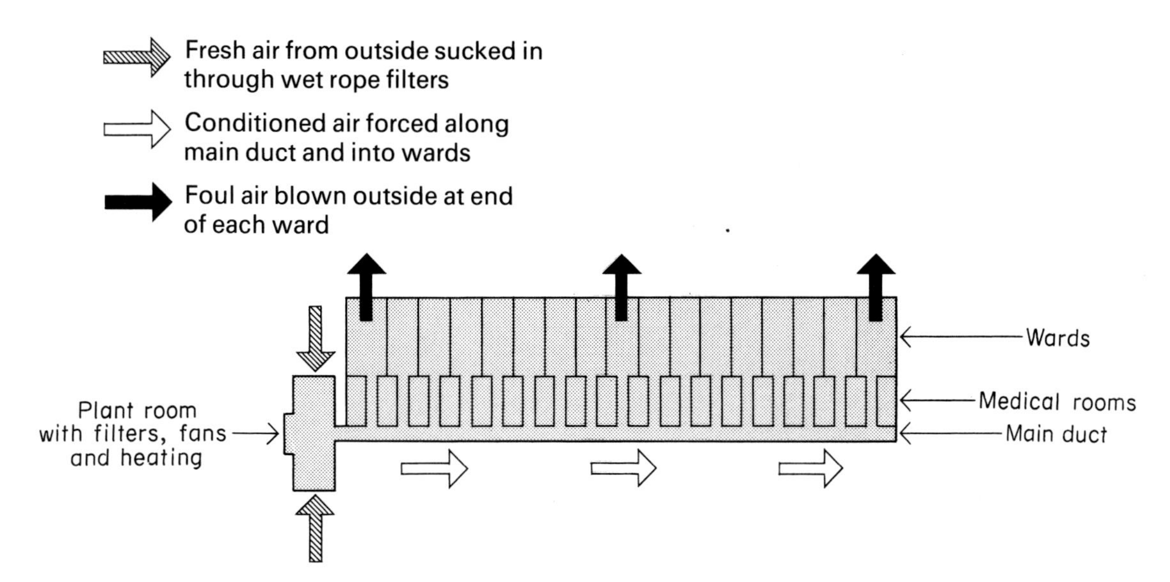

20.4 Royal Victoria Hospital, Belfast, 1903

1930s and 1940s the air-conditioning systems had a great deal of moisture to control and this heavy loading of the plant made installations large and costly. But during the 1950s the fluorescent tube came into general use. It used less current and gave out less heat and so air-conditioning loads were less.

Ventilation fans could be driven by powerful electric motors by 1900 and when air conditioning was installed in American skyscrapers the fans could be located anywhere in the building. This gave designers great freedom for their layout of office floors. In 1949 the American George Bailey worked out that a rectangular-shaped office tower in Chicago with full air conditioning, fluorescent lighting and sound-absorbing ceilings would only cost 8% more than the earlier U-shaped buildings. These earlier types had used windows along the sides of the U-shape for ventilation and had many rooms inside the U-shape which were dark and so difficult to let. During the 1950s the air-conditioned rectangular tower with lifts, escape stairs and vertical ducts in the middle became the common type of office building all over the world.

During the 1950s people began to expect better levels of comfort in every sort of building. The old standards of heating, ventilation and lighting were no longer good enough. More and more mechanical equipment went into buildings. By the 1960s it was not unusual for these mechanical services to be 33% of a building's total cost and by the 1980s this went up to 50%. And it was these costs of mechanical and electrical installations such as air conditioning and power, lighting, telecommunications and computer services and the lifts and escalators and the fire sprinkler systems and the water supply and drainage that led to the High Tech style.

In the same way that people were beginning to dislike high-rise blocks for housing during the 1960s so they were also beginning to dislike high-rise office buildings. At that time the rectangular blocks had continuous outer walls of glass and aluminium panels designed as grids of rectangles. These factory-made **curtain walls** were a type of system building and so were made of endlessly repeated aluminium sections slotted and clipped together in endlessly repeated rectangles. At first the novelty of these gleaming curtain walls was attractive. But as more and more such towers went up in the place of demolished older buildings people saw how characterless city centres had become. All new buildings looked alike. Indeed, all cities began to look alike. (*Figure 20.5*.)

One attempt to restore character to buildings was the rugged concrete shapes of the **New Brutalist style**. But again after the novelty had worn off it too lacked interest. By the 1970s many architects had begun to design buildings with traditional features such as half circle arches and pitched roofs and were using decoration and mouldings. Some housing schemes in built-up areas were almost like the Victorian or Edwardian terraces and squares they replaced.

But also in the 1970s other architects saw that the mechanical services could be used to restore character to buildings. Machinery had become a vital part of buildings and took up a large part of the total cost. Yet much of the machinery was still hidden away behind false ceilings and wall panels. Buildings still gave no clue as to the location of their machine rooms and fan rooms. They still gave no sign as to where the vital duct systems were.

But in the 1970s the young architects were those who ten years earlier as students had tried out new ideas which made features of the mechanical services. From

20.5 Castrol House. One of London's curtain-wall office towers built in 1960

the days of the International style in the 1920s right through to the 1970s a building's structure had been its main feature. But the new architects of the 1970s made a building's mechanical services its main feature.

So instead of being hidden inside the building or in basements or on flat roofs the machinery, the pipes, the ducts, the lift shafts, the air intakes and exhausts all came out into the open and were hung around the outside of the building in a dazzling display of brightly coloured boxes and tubes.

There was silver and scarlet and royal-blue everywhere. Where a wall managed to peep through the pipe work it could be seen to be made of mirror glass or was a panel of fibreglass or of glass-reinforced plastics. Every part of the building that could be seen was made of man-made materials shaped by industrial processes, and finished and polished and coloured by industrial processes.

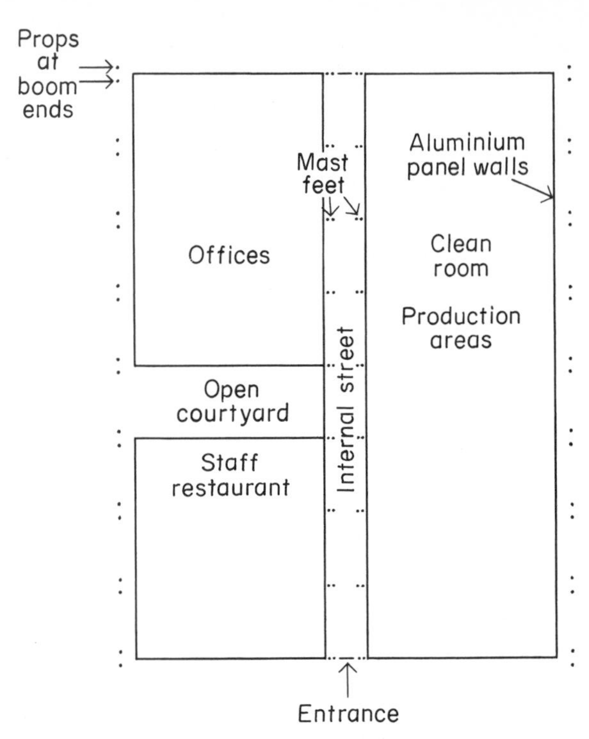

20.6 Plan, Inmos

These striking objects belonged to the world of high technology that was producing spacecraft and satellites and microelectronic circuitry. They were no longer buildings but beautiful machines. The High Tech style had arrived.

At the Inmos micro-chip factory 20 million pounds was spent on equipment for making the chips, another four million pounds on mechanical services and plant and a final four million pounds on the structure to contain the equipment and the services. Inside the production areas dust and other particles in the air had to be filtered out leaving only a few parts per million because dust could ruin the etching of the chips. (*Figures 20.6 to 20.11.*)

Powerful acids and solvents did the etching. Production was continuous so that the factory's toxic wastes were not stored but neutralised and passed into the public sewers. Enormous quantities of chemicals and five megawatts of electric power were needed by the chip-making processes.

The 7422 sq m (80,000 sq ft) of floor space was housed in High Tech rooms with walls of thin double-membrane aluminium panels interchangeable with squares of double glazing. The flat roof had five layers of material with sheet steel in the middle and insulation and waterproofing on top and ceiling finishes and support for services such as lighting underneath. The air-conditioned production areas were serviced by

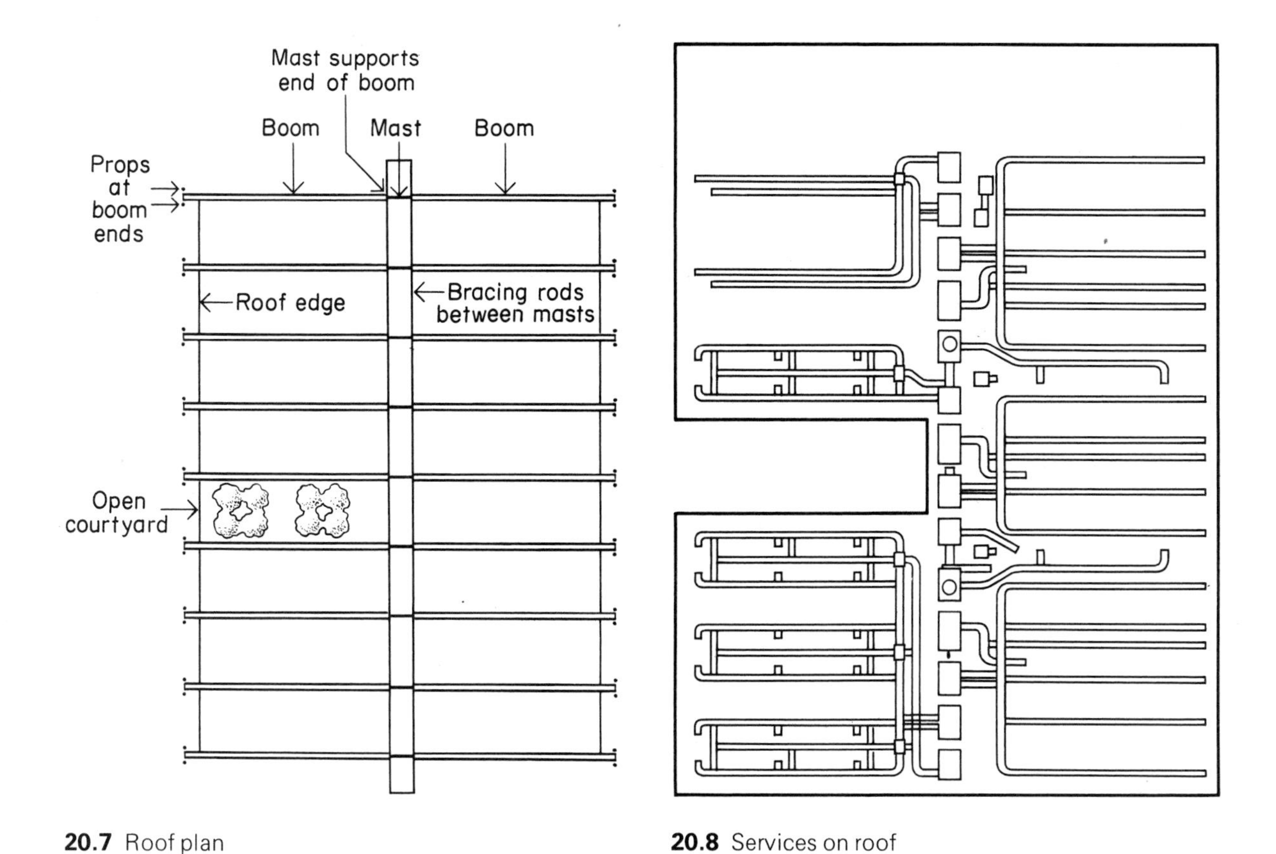

20.7 Roof plan

20.8 Services on roof

20.9 Section of services

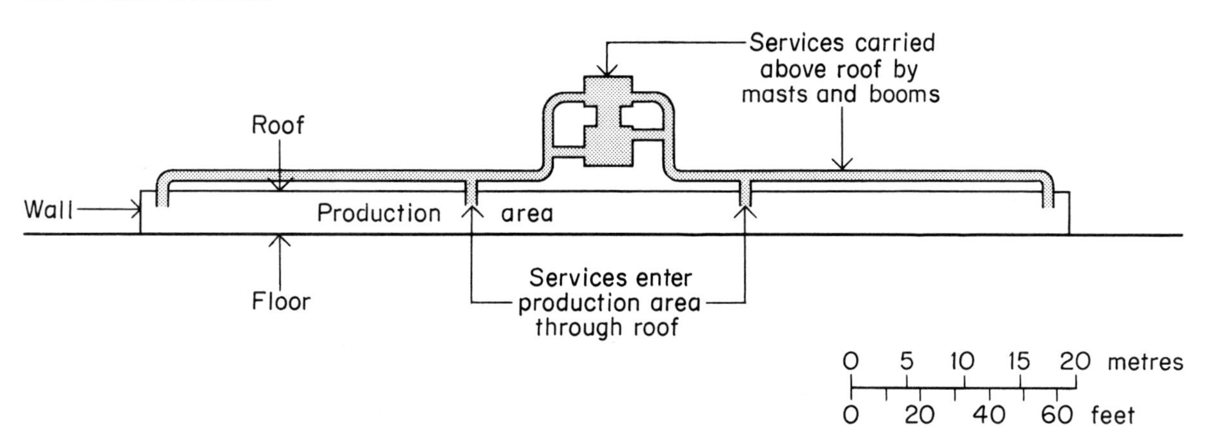

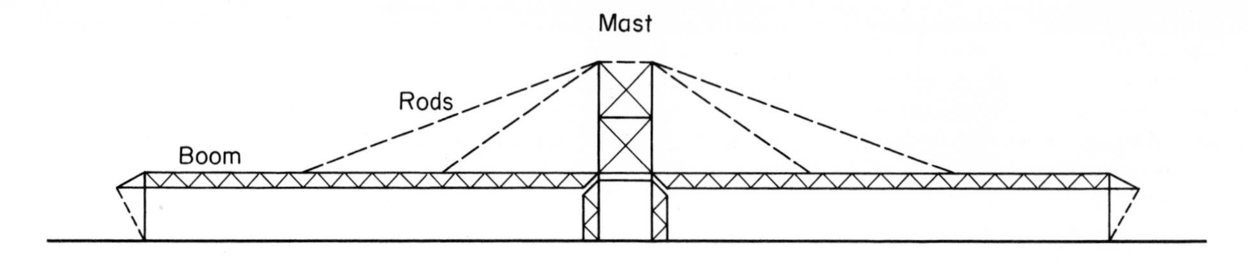

20.10 Section of steel structure

20.11 Detail of steel structure

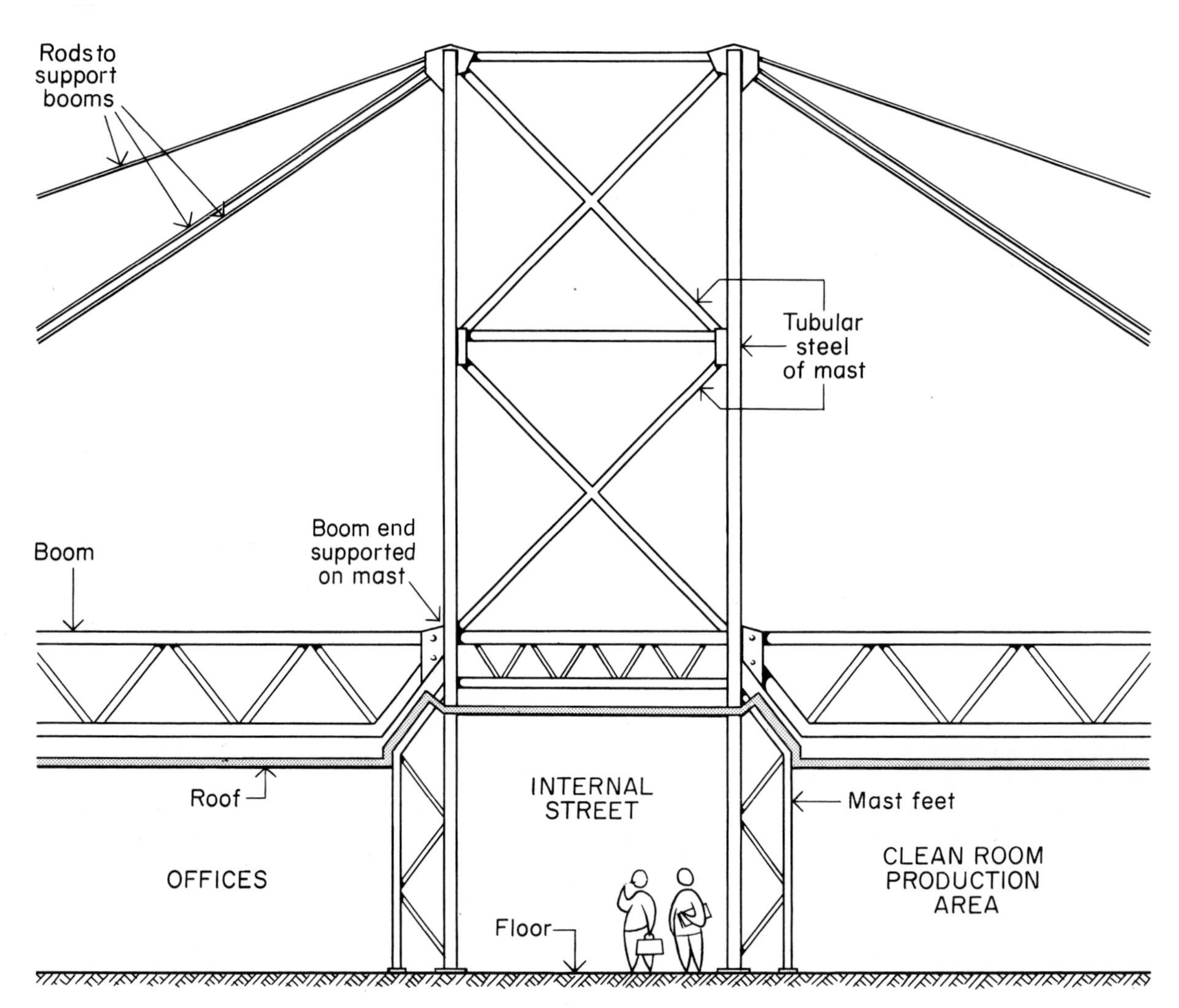

systems up to nuclear reactor standards of performance and safety.

Inmos wanted totally clear floor space in the production areas allowing machinery and equipment to be moved around in new layouts. Service pipes and ducts had to come down from above. The simplest place for all the mechanical services and ducts and the hundreds of pipe runs was above the production area roof. The services had then simply to go directly from the plant room, run along the roof, and down into the production area. An exposed steel structure could support all these services. The structure was arranged in eight pairs of bays, each bay measuring 13 m × 36 m ($42\frac{1}{2}$ ft × 118 ft). This simple, functional answer to the design problem produced a building with all the main features of the High Tech style.

A row of lattice steel masts carried the boxes containing the service plant. The masts were 7·2 m (23·6 ft) wide and were placed 13 m (42 ft) apart. Each side of the masts also carried the ends of lattice beams which were 36 m (118 ft) long. Each mast also supported one end of a 39 m (128 ft) long lattice beam each side. These beams were like the booms of a crane on a building site and they supported the roof of the production areas and the hundreds of pipes and ducts. Each boom got extra support from the mast by two steel ties.

The production areas with their silver-grey aluminium walls sat as a separate structure under the blue-painted steelwork. The building was as striking to look at as a Big Top circus tent going up. Perhaps after all there is nothing very new in building. Tents were a type of structure already old when Stonehenge itself was begun. But tomorrow's tents may be put up by robots.

Further reading

General

Architecture and Technological Change, edited by Pedro Guedes, Macmillan, 1979

A History of Technology, edited by Charles Singer and others, OUP, 1957

Builders and Building Workers, P W Kingford, Arnold, 1973

The Story of Western Architecture, Bill Risebero, Herbert Press, 1979

The Architecture of Britain, Doreen Yarwood, Batsford, 1976

A Guide to Architectural Styles, Herbert Pothorn, Phaidon, 1983

Stone Circles

Stonehenge – Archaeology and Interpretation, R J C Atkinson, Penguin Books, 1979

Stonehenge, Theo Bergström and Lance Vatcher, Bergstrom and Boyle, 1974

The Ancient Stones of Scotland, W Douglas Simpson, Hale, 1965

Fishbourne: A Roman Palace and its Garden, Barry Cunliffe, Thames & Hudson, 1971

The Roman Villa in Britain, edited by A L F Rivet, Routledge & Kegan Paul, 1969

The Architecture of the Anglo-Saxons, Eric Fernie, Batsford, 1983

Pointed Arches

The Construction of Gothic Cathedrals, John Fitchen, OUP, 1961

Portrait of Durham Cathedral, G H Cook, Phoenix House, 1948

Portrait of Salisbury Cathedral, G H Cook, Phoenix House, 1949

King's College Chapel, the Story and Renovation, Rodney Tibbs, Dalton, 1970

Victorian Buildings of London 1837–1887, Gavin Stamp and Colin Amery, The Architectural Press, 1980

Timber Building in England, Fred H Crossley, Batsford, 1951

Golden Rectangles

Architecture in Britain 1530–1830, John Summerson, Penguin Books, 1953

The Classical Language of Architecture, John Summerson, Thames & Hudson, 1980

Inigo Jones, John Summerson, Penguin Books, 1983

The Work of Christopher Wren, Geoffrey Beard and Anthony Kersting, Bartholomew, 1982

Rebuilding St Paul's After the Great Fire of London, Jane Lang, OUP, 1956

Edwardian Architecture, Alastair Service, Thames & Hudson, 1977

Iron Rules
'The First Iron Frames', A W Skempton and H R Johnson/*Architectural Review* Volume No. 131 March 1962, The Architectural Press
Early Victorian Architecture in Britain, Henry Russel-Hitchcock, The Architectural Press, 1954
English Architecture, David Watkin, Thames & Hudson, 1979
London's Historic Railway Stations, John Betjeman and John Gay, John Murray, 1972
St Pancras Station, Jack Simmons, Allen & Unwin, 1968
Structure: the Essence of Architecture, Forrest Wilson, Van Nostrand Reinhold, 1983

High Tech
'Inmos In Gwent' *Architectural Review* Volume No. 172 December 1982, The Architectural Press
Emberton, Rosemary Ind, Scolar, 1983
The Architecture of the Well-Tempered Environment, Reyner Banham, The Architectural Press, 1969
Modern Architecture and Design, Bill Risebero, Herbert Press, 1982
Modern Architecture 1851–1945, Kenneth Frampton and Yukio Futagawa, Rizzoli, New York, 1983
Modern Architecture since 1900, William R J Curtis, Phaidon, 1982

Index